MEASUREMENT AND MONITORING

MEASUREMENT AND MONITORING

VYTAUTAS GINIOTIS AND ANTHONY HOPE

 MOMENTUM PRESS

MOMENTUM PRESS, LLC, NEW YORK

Measurement and Monitoring
Copyright © Momentum Press®, LLC, 2014.

First published by Momentum Press®, LLC
222 East 46th Street, New York, NY 10017
www.momentumpress.net

ISBN-13: 978-1-60650-379-9 (print)
ISBN-13: 978-1-60650-381-2 (e-book)

Momentum Press Automation and Control and Mechanical Engineering Collection

DOI: 10.5643/9781606503812

Cover design by Jonathan Pennell
Interior design by Exeter Premedia Services Private Ltd., Chennai, India

10 9 8 7 6 5 4 3 2 1

Printed in the United States of America

ABSTRACT

This book presents the main methods and techniques for measuring and monitoring the accuracy of geometrical parameters of precision Computer Numerically Controlled (CNC) and automated machines, including modern coordinate measuring machines (CMMs). Standard methods and means of testing are discussed, together with methods newly developed and tested by the authors. Various parameters, such as straightness, perpendicularity, flatness, pitch, yaw, roll, and so on, are introduced and the principal processes for measurement of these parameters are explained. Lists and tables of geometrical accuracy parameters, together with diagrams of arrangements for their control and evaluation of measurement results, are added. Special methods and some original new devices for measurement and monitoring are also presented. Information measuring systems, consisting of laser interferometers, photoelectric raster encoders or scales, and so on, are discussed and methods for the measurement and testing of circular scales, length scales, and encoders are included. Particular attention is given to the analysis of ISO written standards of accuracy control, terms and definitions, and methods for evaluation of the measurement results during performance verification. Methods for measuring small lengths, gaps, and distances between two surfaces are also presented. The resolution of measurement remains very high, at least within the range 0.05 μm to 0.005 μm.

The problem of complex accuracy control of machines is discussed and different methods for accuracy control are described. The technical solution for complex measurement using the same kind of machine, as a master or reference machine, together with the laser interferometer and fiber optic links is presented.

Comparators for the accuracy measurement of linear and angular standards are described, and the accuracy characteristics of these standards are investigated. Accuracy improvement systems using machines are described and examples are given showing the suitability of mechatronic methods for high-accuracy correction and measurement arrangements.

The effectiveness of the application of piezoelectric actuators is demonstrated using the construction of a comparator for angular measurements as an example.

Some mechatronic methods for accuracy improvement of multi-coordinate machines are proposed. These methods control the accuracy of the displacement of parts of the machine, of the transducers or the final member of the kinematic chain of the machine, such as the touch-probe or the machine cutting tool. Some categories of errors may be improved by numerical control means, because they are determined in the form of a graph with some peaks along the measuring axis running off accuracy limits. These points, as numerical values, are provided as inputs to the control device or to the display unit of the transducer and can be corrected at appropriate points of displacement. The idea is introduced for accomplishing the correction by calculating the correctional coefficients in all coordinate directions and the correctional displacement to be performed using the last (conclusive) part of the machine. For example, the last mentioned part may be the grip of the arm of an industrial robot, the touch-trigger probe of a CMM (measuring robot), the cutting instrument or the holder of metal cutting tools, and so on. In this case piezoelectric plates, in the form of cylindrical or spherical bodies, may be incorporated into the last machine member for this purpose.

The experience gained by the authors working at industrial plants and universities, performing EU research projects and international RTD projects, is used throughout the book.

KEYWORDS

monitoring, measurement, machine monitoring, calibration, performance verification, coordinate measuring machines, linear scales and transducers, circular scales, nano-displacement

Contents

LIST OF FIGURES

ACKNOWLEDGMENT

The authors gratefully acknowledge the helpful advice and support provided during the preparation of the manuscript by Dr. Mindaugas Rybokas, Vice-Dean, Faculty of Fundamental Sciences, Vilnius Gediminas Technical University, Lithuania.

ABOUT THE AUTHORS

Vytautas Giniotis

Vytautas Giniotis was employed as Chief Scientific Worker in the Institute of Geodesy, Vilnius Gediminas Technical University, Lithuania, from 2003 until his tragic sudden death in 2012. He had a most distinguished career lasting over 50 years in both industry and academia. His major scientific interests were measurements and instrumentation, 3D measurements, mechatronics in metrology, research into the accuracy of linear and angular transducers, GPS accuracy investigation, and instrument calibration.

He graduated as Dipl. Mechanical Engineer from Kaunas Polytechnic Institute in 1961 and obtained his PhD from the same institution in 1975. His industrial experience covered a period of over 30 years and included a five-year term as Chief Metrologist and Head of the Metrology and Quality Control Department at Vilnius State Grinding Machines Factory between 1989 and 1994. He joined Vilnius Gediminas Technical University in 1994, was awarded Doctor Habilitus in 1996 and appointed as a Professor in 1998.

He is the author of one monograph and the co-author of another one. He has published more than 220 scientific papers in peer-reviewed international journals and international conferences worldwide. He has also published three booklets in his field and has 55 patents and inventions. He was Lithuania's State Science Award winner in 2009 and was an expert member of the Lithuanian Science Council between 2003 and 2008. He was actively involved in collaborative research projects funded by the European Union and was also an Expert Evaluator of European research proposals submitted to the European Commission in Brussels.

Anthony Hope

Anthony Hope is currently Professor of Automation and Control at Southampton Solent University, United Kingdom. He has had a long career in both industry and academia and has worked at Southampton Solent

University since 1990. His major scientific interests include condition monitoring of machines, performance verification of machine tools and coordinate measuring machines, instrumentation and control, materials testing and evaluation.

He graduated from Leeds University in 1964 with a BSc (1st Honours) in Fuel Technology and then spent several years in industry working on the design and development of aircraft fuel systems. He joined Birmingham Polytechnic in 1970 as a Lecturer in Mechanical Engineering and was then awarded his PhD in Mechanical Engineering from Birmingham University in 1976. He has extensive consultancy and research experience in condition monitoring, engineering materials, machine tool calibration, nondestructive testing, and instrumentation and control. He is a Chartered Engineer, an Honorary Fellow of the British Institute of Non-Destructive Testing and a Foundation Fellow of the International Society of Engineering Asset Management.

He is the co-author of one textbook, *Engineering Measurements*, and has written chapter contributions in two further textbooks. He has also published well over 100 scientific papers in technical journals and refereed international conference proceedings. He has collaborated in a number of major European research projects and was awarded the condition monitoring and diagnostic technology (COMADIT) prize in 2008 by the British Institute of Non-Destructive Testing for his contribution to the benefit of industry and society through research and development in condition monitoring.

CHAPTER 1

MONITORING OF PROCESSES, SYSTEMS, AND EQUIPMENT

In this chapter, the methods and means for the monitoring of technical parameters of machines are presented. The mutual interaction between the measurement and monitoring procedures is discussed, and the different levels of monitoring the parameters of a machine are considered in terms of the global, local, and component levels. The importance of sampling procedures is introduced and various sampling strategies for the dimensional measurement of the geometric features of coordinate measuring machines (CMMs) and other machines are discussed.

The importance of determining the information quantity on an assessed object is illustrated; thus providing more complete measurement during the calibration process. Systematic errors and uncertainty in the calibration of linear and circular raster scales are included and probability theory provides a statistical means for evaluating the results of measurement by selecting the pitch of measurement, assessing the set of trials, calculating the mean value of estimates, and evaluating the dispersion at the probability level chosen. The general process presented here provides information on the conditions of measurement performed, as it includes sampling, together with systematic error and uncertainty evaluation.

1.1   INTRODUCTION

The *Concise Encyclopedia of Condition Monitoring* [1] gives an explanation for the main terms and expressions used for the monitoring process. On the basis of the Concise Encyclopedia, system monitoring could be expanded to include a wider scope of the monitoring process. It would include features covering both an upper and a lower level of monitoring.

It can be treated in terms of *global, local, and component* levels of monitoring. The global level includes space and world phenomena, including space satellites and probes, earth and planet investigations, atmosphere phenomena, and environmental processes. An industrial monitoring process can be designated to the global level, followed by the system level, plant level, and machine and equipment level as the components of analysis.

Some authors [2,3] present the analysis of configuring a measurement and monitoring system in the construction, machine engineering industry, and other branches of industry and social activities. The monitoring systems are analyzed to detect the failure of the machine, control the level of vibration, leakage in a water supply network, environment contamination level, and so on. A number of sensors are used within the system for monitoring.

The methodology for measurement system configuration is based on the choice of the system's parameters to be controlled, the limits of the parameters, sample selection, development of the measurement and/or control system, including selection of sensors, selection of information, processing of information, data transfer, and decision making. A general diagram showing the mutual interaction between the measurement and monitoring procedures is shown in Figure 1.1.

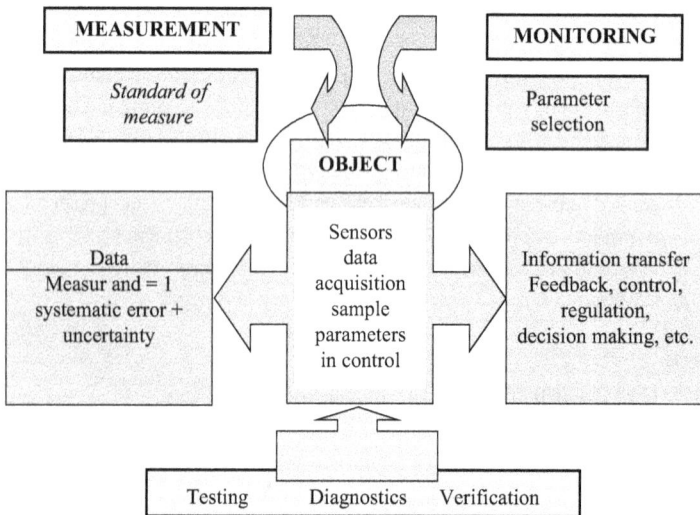

Figure 1.1. General diagram of mutual interaction of measurement and monitoring procedures.

The main difference between measurement and monitoring is the value of the parameter under consideration that is controlled. For measurement the standard of measurement (etalon) is needed for comparison with the readings taken from the object by sensors and the measurement result (measurand) is determined using mathematical/statistical methods while for the monitoring process it is enough to have a predetermined value (desired value) of the parameter to be controlled and to present information about its current value.

This approach can be expanded to the wider scope of global analysis. Here a human activity must be included and investigated such as organization, changes to personal behavior features, social conditions, and economic and financial operations. Then come industrial processes, such as IT organization, environment, procurement, design and documentation, manufacturing and its preparation processes, transportation, marketing, and waste problems and its realization. This global, local, and component level approach can be applied for almost any object being monitored. Every enterprise can be treated as a system consisting of people, machines, and investments having a purpose to produce a quality product. This system is a part of a bigger one, expanding to consortium, national or international community. There we give a definition *global* that in principle is similar to the term *system*.

The global level for machine monitoring would be the surroundings where the machine is placed, connected, and operated. It would include the floor of the plant, the mains supply, and the surrounding atmosphere. The following are examples of features and parameters that would be designated to the global level, for example, shocks and vibrations transmitted through the ground, temperature, humidity and pressure in the surrounding atmosphere, supply of materials and instruments, blank components, operation personnel, the outcome of the production process, software supply and updating, accuracy, and quality control.

The following features and parameters can be designated to the local level, for example, the geometric parameters, quality and accuracy of the basic parts of the machine, drives and motors, control unit, slide ways and datum forming parts, lubrication, sensors, and the coordinate displacement measurement system. Special geometrical specifications can also be included, such as straightness, squareness, run-off of the rotating parts, flatness, perpendicularity of movement of the machine's units, and so on.

The component level must be treated as the parameters that are important to the very component in consideration, such as quality of material, hardness, smoothness, residual stresses, chemical composition, thermal treatment, and so on.

1.2   PARAMETERS FOR SYSTEM MONITORING

The purpose of monitoring and technical diagnostics is an application of all the technical and organization aspects and measurements covering the complete system to ensure a smooth production of goods with appropriate quality in appropriate quantity, by reliable and safe operation of the machines and production components. Some parameters are usually defined for an assessment of the monitoring function, such as mean time between failure (MTBF); mean time between maintenance (MTBM); mean time necessary to repair (MTTR), and some others. It is obvious that when MTBF is to be kept as long as possible, the MTTR is to be scheduled as short as possible.

The machine tool industry has constantly developed more modern forms and structures, beginning from single-station machines, then to computer numerically controlled (CNC) machines, flexible working centers, flexible and robotized production lines, and so on. The same is pertinent for the *Just in Time* system of production as it must be followed by the *Just in Time* system of monitoring. Some technical references have claimed that a planned maintenance system preceded by an appropriate monitoring system increases the efficiency of the production process by up to 20 percent.

The measurement process as described earlier needs more technical, legal, and scientific resources and time. Monitoring is a perfect solution for users to track the machine's health, set-up time, cycle time, part counts, maintenance time, or other productive and nonproductive times. Influences on the upper level have an impact on the machine health, for example, shock and machine vibration are critical to every plant's production accuracy and reliability. The machine condition monitoring and machine test facilities, in-plant or factory-test focused, provide the testing possibilities, including on-line and off-line analyses, high-speed result presentation, and feedback to eliminate errors in manual or automatic mode of operation. The modern customer uses the server architectures, data acquisition, data analysis, and data presentation that are concentrated in one platform of monitoring. Figure 1.2 shows the proposed monitoring levels in relation to monitoring the parameters of a metal cutting machine.

1.3   RANDOM AND REGULAR MODES
OF MONITORING

There is currently a great concern about the potential global warming effect on the Earth. As with most practical theories, it is based on wide

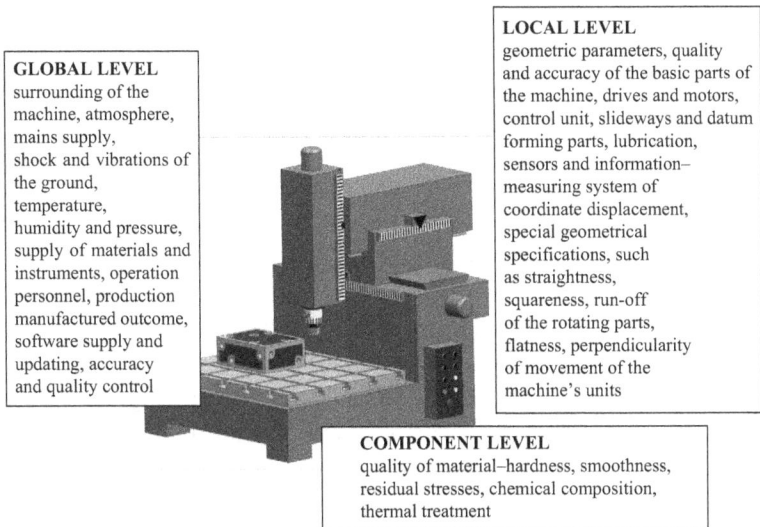

GLOBAL LEVEL
surrounding of the
machine, atmosphere,
mains supply,
shock and vibrations of
the ground,
temperature,
humidity and pressure,
supply of materials and
instruments, operation
personnel, production
manufactured outcome,
software supply and
updating, accuracy
and quality control

LOCAL LEVEL
geometric parameters, quality
and accuracy of the basic parts of
the machine, drives and motors,
control unit, slideways and datum
forming parts, lubrication,
sensors and information–
measuring system of
coordinate displacement,
special geometrical
specifications, such
as straightness,
squareness, run-off
of the rotating parts,
flatness, perpendicularity
of movement of the
machine's units

COMPONENT LEVEL
quality of material–hardness, smoothness,
residual stresses, chemical composition,
thermal treatment

Figure 1.2. Levels of monitoring the parameters of a metal cutting machine.

global warming monitoring. It is generally agreed that global warming has risen by up to 0.6°C on average during the last century. It is understood that there were about 100 temperature-monitoring stations in the world at the end of the 19th century. Nowadays this number exceeds 10,000 and as a result some additional statistical and information data assessment is required to present the information. The same problems arise in many investigations where authors express a need to add more information to the results of measurement giving more data from which to assess the attributes that make, or can make, an influence on the results achieved.

Sampling procedures [4–7] are important in measurement and a typical example is where research has been conducted for comparative evaluation studies of European methods for sampling and sample preparation of soil. Soil analysis has been undertaken to investigate the presence of several metals, and the combined uncertainty has been estimated from the results received. It is important to compare the soil composition from different countries and different places, that is, "outside laboratories" where conditions differ significantly. The process is divided into sampling in the field, thermal and mechanical sample preparation, and laboratory analysis. All this must be determined as appropriately as possible for the purpose of repeatability, reproducibility of conditions and samples, as well as for traceability of the results. The combined uncertainty of the measurement result for 95 percent level of confidence is presented by an expression consisting of eight factors. Special sampling strategies have been proposed, including a "W-pattern" shape of locations from which samples are collected, a spiral trajectory selection for collecting data on a surface, or other patterns of sampling. The aim of all strategies is to cover the data distributed on the surface, or in the informational space, with the highest probability of data acquisition.

Sampling procedures are important in dimensional measurement, as well as the measurement of geometrical features and positions between them. Probability theory is used to provide a statistical means for evaluating the results of measurement (the measurand) by assessing a set of trials, the mean value of estimates, and the evaluation of dispersion at the probability level chosen. It is the basis for data processing in all measurements and is widely used in all kinds and branches of metrology. During the calibration, it is possible to determine only a restricted number of values. An example of the results shows that during the accuracy calibration some larger values of the error can be omitted including significant ones. Therefore, it is important to determine the information quantity on an object that has been evaluated providing a more complete measurement during the calibration processes.

A graphical presentation of a surface curvature (Figure 1.3) shows a difference in curvature when it is represented by the points selected in the top picture (I) from when it is represented by the points selected in the

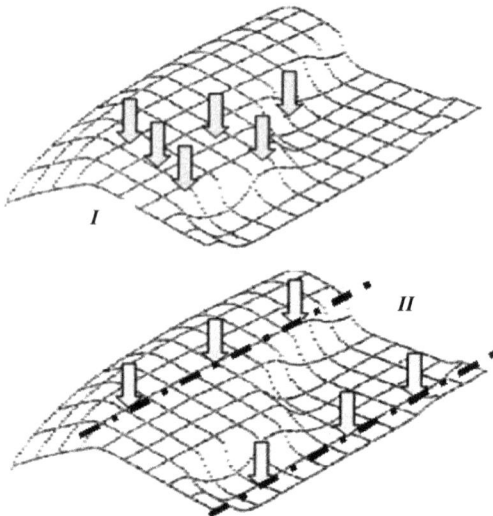

Figure 1.3. Different points selected on the surface surveyed present a different surface form.

lower picture (II). The curvature represented by the points in the lower picture is, in fact, a straight line. A similar situation can be illustrated by the measurement evaluation of roundness by the points shown in Figure 1.4. It is evident that sampling of the surface at the black arrow points would give a much smaller overall diameter than the one sampled by the gray arrow sampling points. It should be observed that the successful solution of the problems mentioned above depends largely on the regularity of the deviation distribution.

It is evident from the research discussed above that a new approach to the sampling strategy and the information quantity derived from the measurement process must be researched further and implemented for scientific and industrial uses. An additional advantage of this approach would give an indication of the sampling value in the measurement result. The second step would then be the efficient use of data presented for the calibration of the object, for data comparison and for achieving the assurance of traceability in the measurements.

Some ideas have been expressed to include the sampling procedure into the equation for the measurand. It is especially important, bearing in mind that modern measuring systems consist of smart transducers that can combine a wide range of data or measurement values, reaching tens of thousands of numerical values. This is also important in view of the traceability of the measurement and other evaluation processes, as it can clearly indicate which part of the information was assessed during the sampling and parameters' determination operations. The quantity of information can be evaluated by joining it with the general expression of the measurement result, that is, expressing the systematic part of the result, the uncertainty of the assessment, and adding to it the quantity of information entropy that shows the indeterminacy of the result.

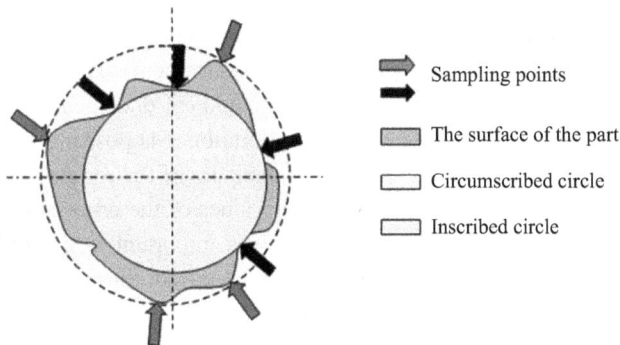

Figure 1.4. Sampling variation example in roundness measurements.

A problem exists due to the large amount of information that is obtained in the calibration of the total volume of numerically controlled machines, such as CMMs. It is technically difficult and economically demanding to calibrate the enormous number of points available, for example, the 324,000 steps of the rotary table of such a machine in the measuring volume arising from six rotary axes, or millions of parts of a meter for a laser interferometer. Hence suitable sampling of measuring points on the surface of industrial parts has been shown to be a very important task, as has the sampling of the points in the machine's volume during its calibration. The task is made even more difficult by the requirement that the accuracy of calibration must remain unchanged, while the time spent must be minimized.

Technical measuring equipment, instruments, and systems, such as linear and circular scales, measuring transducers, and numerically controlled machine tools, provide information about the position of the part of a machine or instrument. A typical measuring system consists of linear and rotary transducers (encoders) whose measuring part is fixed to the moving part of the machine and the index part, the measuring head, is fixed to the base of the machine. Using such systems, it is complicated to measure long strokes of the machine, as the accuracy for a long translational measurement transducer is much more difficult to achieve in comparison with a short one. Accuracy characteristics can be analyzed in terms of informational features of the short measuring transducer (for example, 400 mm) comparing it with the long one (for example, 2000 mm). Usually, modern machines have a resolution of their measuring systems equal to 1 μm, so in both cases the indicating unit of the machine will show 4×10^5 or 2×10^6 digits. Examples of such systems would be the measuring systems of laser interferometers or the display units of CMMs.

Measuring systems are calibrated against the reference standards of measurement comparing their accuracy at some pitch of calibration, for example, at the beginning, middle point and the end of the stroke, or at every tenth of the stroke or range of measurement in the total length or in the circle circumference (for angular displacement). The method used depends on the written standards and methodical documentation of the machines or instruments. During the calibration it is possible to find out only a restricted number of values. An example of the results shows that during accuracy calibration some larger values of the error can be omitted, including significant ones. Hence, it is important to determine the information quantity on an object that was assessed; thus providing more complete measurement during the calibration processes.

The accuracy of linear measuring transducers is checked (calibrated) against the reference standards of measurement (etalons) comparing their accuracy at every 0.1 length of the stroke or alternatively only at the beginning, middle, and the end of the measuring stroke. Some research has been undertaken to determine the distribution of errors in the measuring length of linear measuring transducers. The intervals of verification of the accuracy of linear transducers have been checked against the reference measurement by changing the length of the intervals. The digital output of photoelectric translational and rotary transducers have the last digit number equal to value of 0.1 μm or 0.1″ (seconds of arc). The measuring range in these cases is equal to 10 to 30 m, and a full or several rotations of the shaft to be measured. The value in arc seconds of one revolution is 1,296,000″, that is, 12,960,000 discrete values in the display unit. It increases more than 10 times if the indication is to be at every 0.1 of the value. So, the measurement results in the display unit can be proved by metrological means only at every 100, 1000, or a similar number of strokes. Information entropy allows us to present to the user which part of the information available from the measuring device is assessed by metrological calibration means. So, the analysis presented here is concerned with the information evaluation of the measuring system of the machines. Practically all the moving parts of the machine together with its information-measuring systems are involved in the manufacturing of a component on this machine. Therefore, all inaccuracies of the information measurement system translate into the inaccuracies of the part produced. Information entropy provides an approach to reach the cognitive information about the accuracy available in the system under consideration.

Nevertheless, there is little research performed up to date to join the terms of uncertainty and indeterminacy in the result of data assessment. It would be especially important for the assessment of measurement data. Measurement data are assessed according to mathematic statistics permitting us to see the result evaluated by the systematic part of the measurand and the uncertainty according to the probability level chosen. The data on the sample taken into account during the assessment procedure would give very useful information in all data assessment processes.

The connection of *uncertainty* with *indeterminacy* would provide a large additional informational value to conceive the comprehensiveness of an analysis performed. It will be connected with a common approach to traceability with the low level of uncertainty changing the methods and standards for calibration services. The proposal should result in a reduction of costs, possibly up to 30 percent, for measurements and calibration. The introduction of new type three-dimensional (3D) measurements would affect manufacturing industries in avionics, electronics, machine engineering, civil engineering, and military facilities. It would cover future needs for a larger scale of implementation and it would strengthen the position of European manufacturing industry in the market place.

The sampling strategy for dimensional measurement of geometric features of CMMs and other machines has a significant role in the machines' accuracy testing and verification. The Hammersley sequence [8,9] and a stratified sampling method are used for the assessment of geometric accuracy of the workpiece depending on its manufacturing process and dimensional accuracy. The objective is to achieve the necessary precision of the estimation of the dimension's value at minimum cost; it is seeking a better efficiency of the measurement process. The Hammersley sequence is effective for determining the proper sampling size for each geometric feature and can be compared with other methods such as the uniform sampling and random sampling methods. The scientific analysis performed shows that the sampling strategy based on the Hammersley sequence gives a nearly quadratic reduction in the number of samples compared with the other sampling methods. It is stated that the Hammersley sequence method saves time and cost while maintaining the same level of accuracy. The sampling strategy for various geometric features of part measurements in industry is especially important in computer-integrated manufacturing, where both time and cost are significant.

The determination of a systematic error of the CMM in its measuring volume can be performed by applying the *L-P* sequences method during the calibration of the CMM. This method enables the selection of a large amount of information about an accuracy parameter using quite a small number of trials. The aim of this method is to minimize the number of points and strokes of measurement during the coordinate and geometric error assessment of CMMs and to collect the optimum amount of information. Preliminary research has been made to determine the correlation between the geometric errors measured at various distances and steps in the working volume of the machine. The research has shown that strong correlation between the geometrical accuracy parameters exists, the correlation coefficient being 0.7 to 0.8. Then steps for reducing the number of points to be checked have been applied. The method of *L-P* sequences has been used to find the points equally distributed in the working volume. Evenly distributed points according to the principle of *L-P* sequences are displayed in Figure 1.5.

The points are distributed most evenly in the case when the number of points is $N = 2^m - 1$, $m = 1, 2, ...$

The coordinates of *n*-points in the *L-P* distribution are

$$Q_i = (q_{i1}, ..., q_{in}), \qquad i = 1, 2, ... \qquad (1.1)$$

(a)   (b)

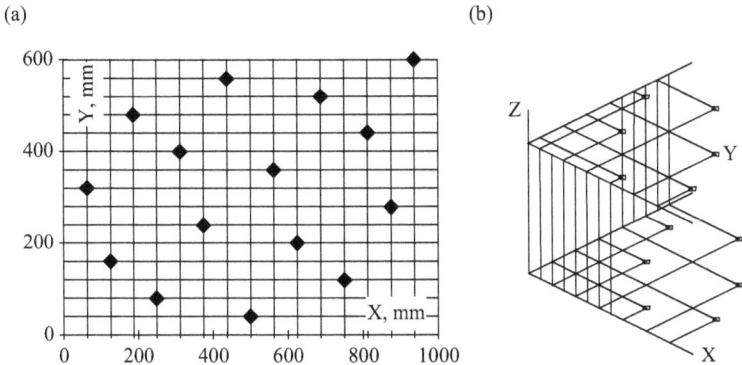

Figure 1.5. The points of a *L-P* sequence generated in plane (a) and in the volume (b).

i—the index of a point. By writing the indices of points $i = e_m,...e_1$ in the binary system and after performing a logic operation, it will become

$$q_{ij} = e_1 v_j * e_2 v_j^2 * ... * e_m v_j^m; \qquad (1.2)$$

The operation designated by * performs the comparison of codes. The values of v_j^s are taken from the tables whose size is $1 \le s \le 20$, $\quad 1 \le j \le 51$. Using $n \le 51$, n-dimensional Q points will be found, $N \le 2^{21}$.

The coordinates of points are calculated after determining i for every number: $m = 1 + [\ln i / \ln 2]$.

After that $j = 1, ... , n$ is calculated.

$$q_{i,j} = \sum_{k=1}^{m} 2^{-k+1} \left\{ \frac{1}{2} \sum_{l=k}^{m} [2\{i2^{-e}\}][2\{r_j^{(e)} 2^{-k-1-e}\}] \right\}. \qquad (1.3)$$

Here [ ] indicates an integer, and { } a fraction. A FORTRAN program was used to generate the sequence of integers consisting of 25 points evenly distributed in a three-dimensional cube. The prognosis of declination $\delta X_i / \delta Y_i$ of additional points from the sequence was performed by means of the Lagrange polynomial:

$$\delta X_k = \sum_{i=1}^{n} \delta X_i \prod_{j=1}^{n} \frac{X_m - X_j}{(X_i - X_j)}, \qquad (1.4)$$

where X_m—a point of *L-P* sequence and $m = 1.25$.

An example of the points generated in the plane is presented in Figure 1.5(a). The method helps us to determine the coordinate errors using a smaller number of measurements. It is possible to take only the special steps of measurement and both to save time and ensure a higher accuracy of the procedure as well.

According to the coordinates of points generated using the equations 1.1 to 1.4, an experimental template (of dimensions 420 mm × 620 mm) was made and used for CMM accuracy calibration. The points determined by *L-P* distribution were fixed by holes of 8 mm with a fine machined surface. The tip of the measuring head used was 3 mm diameter. Measurements were performed at two levels in the height: on the surface of the machine's table and on a surface at 200 mm in the *z* direction. The results achieved show no more than approximately 15 percent difference from the results achieved during the calibration performed using conventional methods and means.

The application of *L-P* sequences minimizes the number of points and strokes to be measured during the coordinate and geometric error assessment of automatic machines. Evenly distributed points according to the principle of *L-P* sequences are used for manufacturing the simple template that is effectively applied into the three coordinate directions of the machine to be calibrated or under the verification of its performance.

An efficient sampling strategy for CMM calibration purposes has some other implementations. Applying the *L-P* sequences method for the sampling strategy during the calibration permits the determination of a systematic positioning error of the CMM in the plane or its measuring volume. This method permits the selection of a large amount of information about accuracy parameters using a smaller number of trials.

The method for selecting Hammersley sequences is described as follows:

$$P_i = i / N, Q_i = \sum_{j=0}^{k-1} b_{ij} 2^{-j-1}, \tag{1.5}$$

where N—the number of summary points of selection; $i \in [0, N-1]$; b_i— a binary display of index i; b_{ij}—jth bit of b_i; k—$[\log_2 N]$. With the help of the mathematical relationship mentioned above, the method for selecting points is programmed by Hammersley's algorithm.

The coordinates of 8 points generated by four different methods in the *xy* plane are shown in Figure 1.6. From the examples provided one can see that every method generates different coordinates. Analyzing the mathematical processes of the methods mentioned above one can note that they generate point coordinates without regard to the regularity of the deviations in surfaces or planes being measured. Therefore, different methods generate very different coordinates of the points measured.

(a)

(b)

(c)

(d)

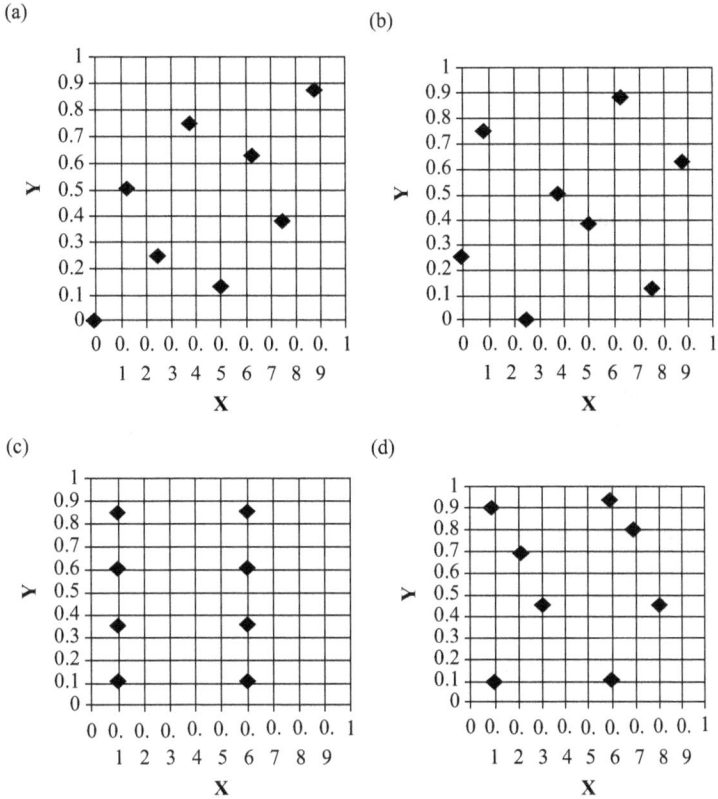

Figure 1.6. The coordinates of the 8 points are generated by different methods in the xy plane: (a)—by Hammersley sequence method; (b)—by Halton–Zaremba sequence method; (c)—by the method of straightened systemic selections; (d)—by minimizing the number of points measured using a certain algorithm.

The selection method proposed by Halton–Zaremba differs from the Hammersley sequence methods; its mathematical model is as follows:

$$P_i = i / N = \sum_{j=0}^{k-1} b_{ij} 2^{-(k-j)}, \qquad (1.6)$$

$$Q_i = \sum_{j=0}^{k-1} b'_{ij} 2^{-j-i}, \qquad (1.7)$$

where N—the number of summary points of selection; $i \in [0, N-1]$; b_{ij}-the jth information bit of b_i; b'_{ij} is $(1 - b_{ij})$ when j is odd and otherwise $b'_{ij} = b_{ij}$; k is $[\log_2 N]$.

Another approach to measurement data sampling, the gray theory method, has been used to determine point coordinates of another product and to predict production deviations of its geometrical parameters. The essence of this method is to generate a new series whose members meet the following requirements:

- Let the first member of the series being generated be a member of the existing series.
- Let the sum of the first and second members of the existing series be the second member of the new series.
- Let the sum of the first, second, and third members of the existing series be the third member of the new series, and so on.
- The new series generated is called the accumulated series of the same time.

The initial series is as follows:

$$x^{(0)} = \{x^{(0)}(1),\ x^{(0)}(2),...,x^{(0)}(n)\}. \tag{1.8}$$

Then the newly generated series of the first time using the technique would be

$$x^{(1)} = \{x^{(1)}(1), x^{(1)}(2),\ ...,x^{(1)}(n)\}, \tag{1.9}$$

where $x^{(1)}(k) = \sum_{i=1}^{k} x^{(0)}$, $k = 1, 2, \ . \ . \ . \ , n$.

A sequence of 30 measurement points generated using the gray theory method is illustrated in Figure 1.7.

Information concerning the accuracy parameters of the coordinate position of the parts of multicoordinate machines, robots, and CMMs is quite complicated for selection and assessment. The great bulk of information in the measuring volume and the lack of a spatial reference measurement are the main technical and metrological tasks to overcome. Some technical methods for accuracy measurement and some new techniques for complex accuracy assessment have been proposed. The method of L-P sequences has been used for the determination of points equally distributed in the working volume. The purpose is to minimize the number of points to be measured during the coordinate error assessment of the machines. In addition, it is important that the information received must not be less than that for assessment of the accuracy of the machine's geometric elements, performed according to conventional methods and means.

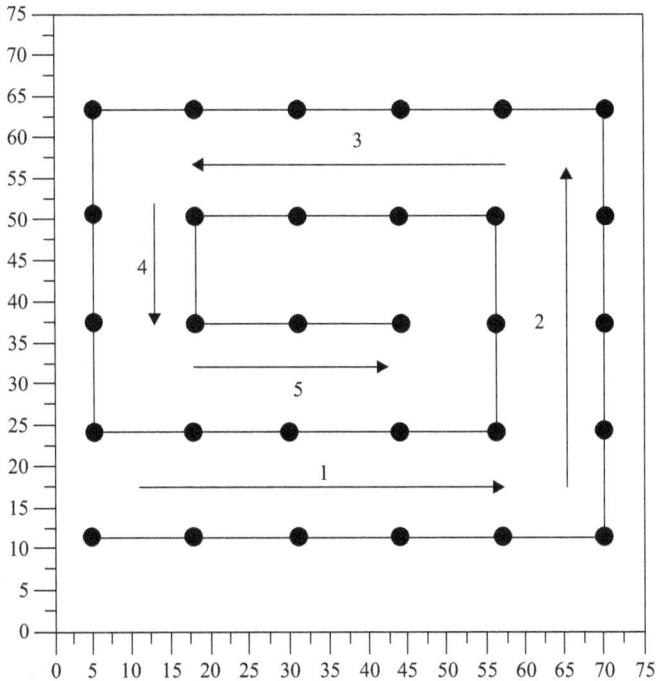

Figure 1.7. A sequence of 30 measurement points generated with the help of the gray theory.

When analyzing the statistical L-P sequences method and the gray theory method the advantages of the L-P sequences method become clear, especially when applying it to 3D measurement. All methods for determining measurement coordinates have advantages compared with the standard methods in that they shorten the measurement time, but the expediency of applying them has to be considered in every practical case separately. The gray theory method, the straightened systemic and systemic random selection methods and those of local sampling are better suited for measuring errors on a plane, while the methods of L-P sequences and others are better suited for measuring errors in 3D. If any *a priori* information on the regularity of deviation distribution is available, it is possible to select a measurement method more precisely and to substantially cut the time of data acquisition.

The method of splines is based on subdividing the surface being measured into certain fields (areas). The physical and mathematical models are created by using the surface characteristics or data measured in these fields or areas. Special splines are used for creating a mathematical model of a surface:

$$C(t) = \sum_{i=0}^{n} N_{i,m}(t)V_i, \qquad (1.10)$$

where n is the number of points controlled, and $V_i (i = 0,1,2,...n)$ is the number of points in a 3D measurement space. The magnitude $N_{i,m}(t)$ is estimated by using the Coxde Boor algorithm:

$$N_{i,1}(t) = \begin{cases} 1 \ (\xi_i \leq t \leq \xi_{i+1}) , \\ 0 \ (\text{otherwise}) , \end{cases} \qquad (1.11)$$

$$N_{i,m}(t) = \frac{t - \xi_i}{\xi_{i+m-1} - \xi_i} N_{i,m-1}(t) + \frac{\xi_{i+m} - t}{\xi_{i+m} - \xi_{i+1}} N_{i+1,m-1}(t). \qquad (1.12)$$

The sequence of magnitudes $\xi_j (j = 0,1,...,m + n)$ is a vector of a node element that satisfies $\xi_j \leq \xi_{j+1}$; the nodes are selected to meet the requirement $0.0 \leq \xi_j \leq 1.0$. Magnitudes $V_i (i = 0, 1,...,n)$ are unknown. They are determined by processing measurement results and applying B-spline curves to them. By applying the techniques presented, a mathematical model of the physical surface model can be obtained.

In CMM calibration it is important to choose the length measuring system as the standard measure to calibrate all three coordinate axes of the CMM. A precision length measuring system was discussed in a paper by Tae Bong Eom and Jin Wan Han in 2001 [10]. A new linear measuring machine was described for the calibration of a variety of length standards using a laser interferometer as a reference measure. Such equipment is very similar to most calibration arrangements used for the calibration of CMMs. In the article the authors show the importance of avoiding the influence of Abbé offset, cosine error, and tilting errors occurring during the calibration. It is important to determine the errors in the different directions of the coordinate axes in the measuring volume. Measurements are performed in the diagonal direction of the measuring volume and this method is efficient because it involves three dimensions in every step of the measurement performed. A redundancy technique was employed in some methods of calibration where the calibration in the area or volume was performed by placing the reference measure at several different positions. An optical scale was used for such purpose. The systematic errors of the machine were identified and after mathematical analysis the errors could be corrected by technical means.

Easy handled free-standing ball bar systems have been widely used to evaluate the volumetric performance of CMMs. The technical requirements and design properties for successful application of ball bar systems have been analyzed. However, the calibration process using such a system was restricted by the few points in the area or the volume that could be checked by the ball bar gauge. The CMM errors have been investigated by applying statistical analysis for the determination of the uncertainty of the CMM. This is an important application having in mind the shortage of measured points during the calibration using a ball bar gauge.

Sladek and Krawczyk [11] highlighted the problem of the large amount of information that needs to be evaluated in the total volume of a large CMM. A 3D artifact consisting of nine spheres was used for the calibration. The authors showed that there is no technical means and economic justification for the calibration of 324 000 steps of the rotary table of a CMM in a measuring volume consisting of six rotary axes.

The suitable sampling of measuring points on the surface of industrial parts is a very important task and the same is relevant for the sampling of points in the measuring volume of the CMM during its calibration. At the same time, the accuracy of calibration must remain at the same level, although the time consumed during the calibration will be shortened. Cheaper servicing of CMMs is required and the expected savings to be gained from this development should help to reduce the comparable cost of CMMs for any given measurement range. Hence, the expected design benefits may not only make the CMM appropriate for a wider range of applications, but also the reduction in cost would make the use of a quality measurement device more readily available to other users (particularly SME companies).

Kim et al. [12] investigated three-dimensional coordinate metrology using a volumetric phase-measuring interferometer. The authors presented three-dimensional coordinate metrology using a three-dimensional movable target and a two-dimensional array of photodetectors. The x, y, and z location of the target was determined in the CMM volume. An important feature of the investigation was the analysis of the computation time, depending on the number of point references and the selection of point references for numerical search. A statistical approach was used to show the repeatability of the measurements and compare them with conventional measurements.

Research on the distribution of errors inside the calibration pitch along the measuring length of a linear translational transducer shows the importance of choosing the interval for the determination of the displacement measuring system's accuracy. Ideally the accuracy of the linear translational transducer, or rotary encoder, should be known at its every output digit. During the calibration it is possible to find out only a restricted number of values. Examples of the results show that during the calibration of the accuracy with a larger pitch, some values of the error can be omitted, including significant ones. Therefore, there is a need to determine the quantity of information on an object that has been estimated providing more complete measurement information during the calibration process.

The digital output of photoelectric rotary and translational transducers has the last digit equal to a value of 0.1 µm or 0.1″ (seconds of arc). The measuring range in these cases is equal to 10 to 30 m in the linear transducer, and a full rotation, or several rotations, of the shaft in the circular transducer. The value in arc seconds of one revolution is 1,296,000″, that is the same number of discrete values in the display unit. It will increase to 10 times this number if the indication is to be at every 0.1 of the value. The measurement results indicated in the display unit can be verified by metrological means (calibrated) only at every increment of 1/100, 1/1000, or even a smaller number of the total information. A suitable theoretical approach to provide more information in this area would be to express the measurement result in terms of the systematic part of the measurement, the uncertainty and the probability level.

Figure 1.8 shows examples of stochastic (a) and periodical (b) modes of sampling. The random values X and Y are to be assessed in the set of a stochastic distribution. It is demonstrated that a random sampling can lead to sample 1 having only the random variable Y approximately with normal distribution, or sample 4 having almost only the random variable X approximately with a biased normal distribution. Quite a different number of variables with different distributions are presented in samples 2, 3, and 5. Such a picture is very possible in real sociological, chemical, and biology investigations. Figure 1.8 (b) is an example of periodical structure that is more relevant and characteristic for technical information systems with a usually normal distribution law.

There is plenty of research on sampling strategy in mechanical engineering and the previous example which deals with the metrology of circular shapes of manufactured parts by measuring a number of sampled points on the surface can be applied in other fields in the evaluation of geometrical parameters on CMMs.

The result of measurement is strongly influenced by the number of points selected on the circumference of the surface and their distribution. The sampling strategy is analyzed with the aim to evaluate the errors made during the appraisal of the geometric circular feature. It helps to make a

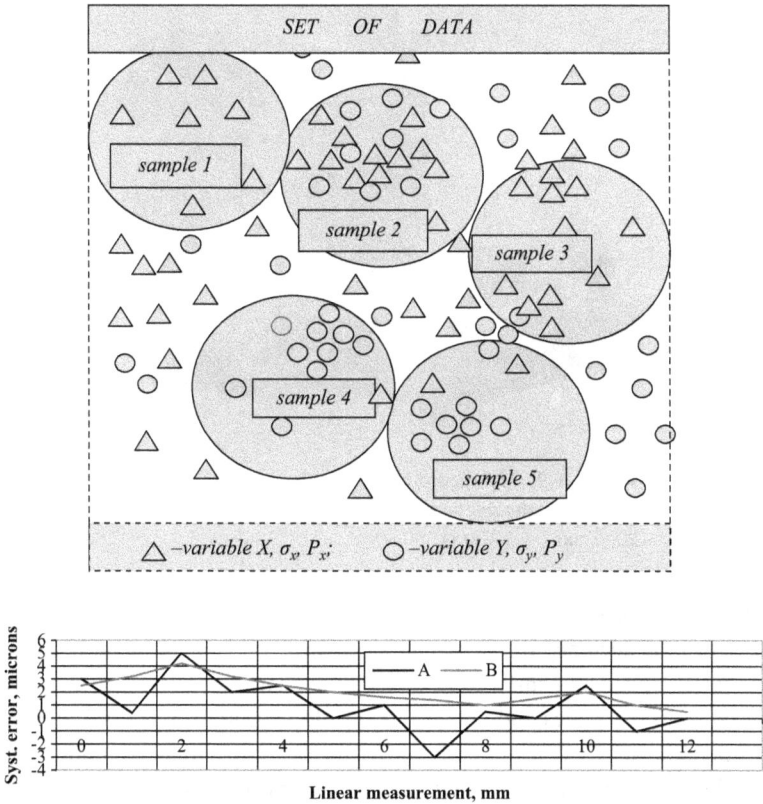

Figure 1.8. Sampling examples in stochastic (a) and regular (periodic, b) modes.

decision as to how many points are necessary for measurement of the circle and how they must be distributed on the surface.

1.4 PRINCIPLES OF PERFORMANCE OF INFORMATION MEASURING SYSTEMS

Linear and circular raster scales are widely used in technical machines and measuring equipment, where they serve as a reference measure for the control of the position along the axis of movement (Figure 1.9). These scales are used in control systems with an optical readout, or they are incorporated into linear or rotary encoders for the determination of the position of a cutting tool or the tip of a touch probe for the measurement of machined parts, for example. The accuracy of the position of the strokes or lines in

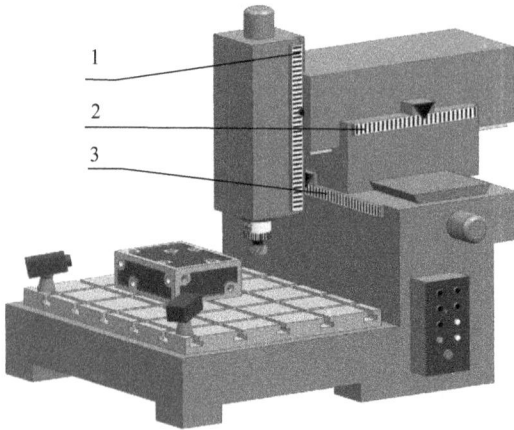

Figure 1.9. Metal cutting machine's measuring systems in the relevant coordinate axis: 1—Z axis, 2—Y axis, 3—X axis.

the raster scale essentially determines the accuracy of the scale and, subsequently, the accuracy of the machine into which it is incorporated. Linear or rotary encoders consist of reading heads that follow the placement of strokes on the scales and convert them into an output digital signal corresponding to a value of displacement in linear or circular motion. The reading head operates in inductive, photoelectric, or magnetic mode, all of them transferring information into a digital output. Output information is presented up to 0.1 μm digits or "0.1" (seconds of arc) values.

There have been many investigations into the operation of the accuracy parameters of machine tools or measuring equipment but less has been published in the field of optical scale position accuracy. At the same time, significant developments in optical sensors, as a means to use fiber optics to assure high accuracy and remote operation in measurements, have occurred. An analysis of the methods and means of calibration of scales shows that there are still some problems to be solved in this area. One of the most important tasks remaining is the enhancement of the flow of information gained from the measurement process. The second task concerns the importance of the sampling process for the performance of the calibration. The measurement of the scales requires a number of points with the pitch of the measurement selected along the full circumference of the circular scale or along the length of the linear scale. Those parameters are closely connected to each other. It is impossible to calibrate each stroke of the raster scale using conventional means of calibration, as there are no standards of measurement for each pitch of the raster scale when the pitch of the strokes, for example, is equal to 20 μm (for the linear scale) or between 1′ and 5′ (minutes of arc for the circular scale). Two different approaches to solving such sample problems are presented.

Linear and circular raster scales are produced using linear and circular dividing machines using modern techniques involving photo lithography and vacuum technology. For a long time mechanical machines were used for this purpose, where mechanical angle standards were used for circular dividing or angular position determination purposes. In most cases, a worm-glass gear was used for circular dividing machines. The accuracy of the angular position fixed by these devices is about 0.3″ to 0.5″. The "run-out" error of rotation of the axis in such machines is within 0.1 μm. A master lead-screw was used in the linear dividing machines for the production of the linear scales. Raster scales, produced by this equipment, serve as the main element for rotary and linear transducers of high accuracy. The transducers are used as a reference measure for metal cutting tools, robotics, and measuring equipment, such as coordinate measuring machines, instruments for navigation purposes, and geodesy, for example.

Errors of angle are determined by several methods, frequently used in machine engineering and instrumentation. In such methods the calibration of the scale may be performed by comparing the strokes of the circular scale with those of a reference scale. The standard angle of measurement, arranged by using precise angular indexing devices or a transducer, is used in other methods. A standard of measurement can also be generated by the use of time pulses, comparing the time intervals with those formed from the strokes of the scale during its rotation.

A wide range of methods for circular scale measurement are known from astronomy and geodesy. Most of them are based on calibration within the full circumference with the angle formed by using measuring microscopes, positioned on the relevant strokes of the scale. These methods are generally known by the names of the scientists who have developed them, such as *Bruns and Wytozhenz* (mathematicians), *Schreiber, Perard, Wielde,* and *Jelisejev* (physics and metrology scientists), for example. In most of the methods, the control angle value is calibrated by comparing the angular pitch between the microscopes with the value of the position of the strokes to be measured. An assessment of the difference between the values is expressed by a mean value and every bias from this value is determined as an error of the stroke of the circular scale. Therefore, "correction" values may be determined as well as errors of the "diameter" of a circular scale. In this case, the term *diameter* means the line going through the opposite strokes of the circular scale. The calibration process varies when using different values of control angles, for example, 10°, 20°, 30°, and so on. The disadvantage of the method is that the errors so determined have different values at different places on the circumference and the graph of the errors is not correlated exactly with the center of the scale. This deficiency is improved by using a method in which the opposite direction of measurement is also included and a different value of control angle is used (e.g., after using 36°, another angle value of 9°, is used). Some other methods show an improvement by expressing the errors through Fourier analysis. Control angles of 36°, 45°, and 60° are used, with the purpose of measuring the scale with a pitch of 3°. Systematic errors and uncertainty in the calibration may be determined and it should be noted that the errors determined are bound to the diameters of the circular scale.

The time taken for measurements using these methods is extremely long. Of course, this is mostly due to the difficulties of automation of the measurement process. Measurements are performed in steps, mostly of 3° or more. Using these methods significant problems of time and loss of accuracy are encountered; an especially difficult task is to measure the scale within a pitch of 1° or less. The raster circular scale can include from several hundred to several tens of thousands of strokes. It is evident that there is no easy means for creating the standard angle for the measurement of each stroke and therefore the errors of the *short period* (within the period of 1° to 3°) are not determined. The methods developed for solving problems of this kind cover a range of optical applications, including the novel usage of the standard angle of the half circle and the new application of Moiré fringe techniques for this purpose.

Research on the distribution of errors inside the calibration pitch along the measuring length of a linear translational transducer has been performed. It is a multivalue measurement as the accuracy of the transducer is required to be known at every output digit. During the calibration it is possible to find out only a restricted number of values. The accuracy of a linear transducer was calibrated by changing the length of the calibration intervals. An example of the results shows that during the calibration of the accuracy with a larger pitch, some values of the error can be omitted, including significant ones. Therefore, there is a need to determine the quantity of information on an object that has been estimated providing more complete measurement information during the calibration process.

The digital output of photoelectric rotary and translational transducers have the last digit equal to a value of 0.1 μm or 0.1″ (seconds of arc). The measuring range in these cases is equal to 10 to 30 m in the linear case, and a full or several rotations of the shaft in the circular case. The value in arc seconds of one revolution is 1,296,000″, that is the same number of discrete values in the display unit. It will increase to 10 times this number if the indication is at every 0.1 of the value. The measurement results indicated in the display unit can be proved by metrological means (calibrated) only at every increment of 1/100, 1/1000, or even a smaller number of total information.

For example, the exact position of the strokes in the raster scale, having m strokes (Figure 1.10), is assessed (calibrated) by technical means only at the pitch of calibration—at every k number of strokes. The pitch is chosen in accordance with the accepted global standards and can cover only a small number of strokes of the raster scale or the digital output of a laser interferometer or a similar instrument. It means that the information is checked only over a small part of the scale or the output of the digital instrument. A significant quantity of data is obtained during other mechanical measurements, such as coordinate measurements of parts and flatness measurement. That is why the sampling problem is very widely analyzed in many pieces of research concerned with these problems.

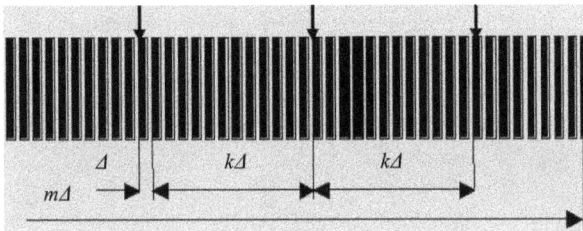

Figure 1.10. Raster scale having calibrated strokes at some pitch.

A similar situation for the calibration of a circular raster scale is shown in Figure 1.11. Here the pitch (30°) is determined by the standard measure, in this case a polygon with 12 angles.

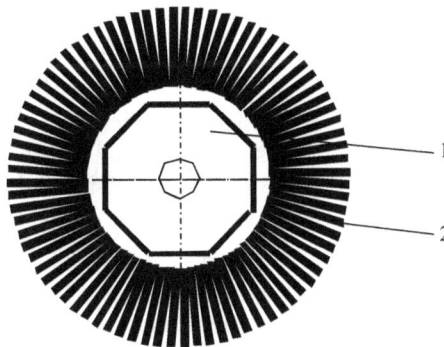

Figure 1.11. Circular raster scale's calibration at the pitch of the standard measure—the polygon with 12 angles: 1—polygon, 2—circular raster scale.

An analysis of the positioning error variations for a linear translational transducer also shows different accuracy values inside the pitch of
calibration. The point at the beginning of the measurements was changed
by 2 mm when performing the calibration of the full length (400 mm)
of the transducer with the pitch of calibration being 10 mm. Mean arithmetic values of the systematic error at every change of the beginning of
the measurements were analyzed. A mean value at a displacement of the
beginning of 2 mm, 4 mm, 6 mm, and so on was determined and these
values were compared between themselves. Empirical dispersions at these
positions have been designated as S_2^2, S_4^2, ... The confidence intervals
have been evaluated by the means of mathematical statistics. The empirical dispersions S_6^2 and S_8^2 are the largest, so they have been checked by
statistical methods. It shows that the evaluations of empirical dispersions
at spatial intervals S_6^2 and S_8^2 belong to the same random value entity. The
values calculated have been obtained by using expressions:

$$G_m = \frac{S_m^2}{S_0^2 + S_2^2 + \cdots + S_{18}^2} = \frac{S_m^2}{\sum\limits_{i=0}^{2k} S_i^2},$$

where $m = 6,8$; $k = 1, 2,...,9$, and $t_m = \dfrac{\overline{\delta l}_m - \overline{\delta l}_k}{\sqrt{S_m/n_m + S_k/n_k}}$,

where $\overline{\delta l}_m$—the mean arithmetic value calculated ($m = 6$; 8); S_m—the estimate of the dispersion in the range of measurement; n_m—the quantity of
measurements; S_k—the estimate of the dispersion in the range of measurement consisting of n_k measurements.

The calculation of G_m produces the results of 0.098 and 0.115. At the
value of $P = 0.99$, the critical value is 0.212. The checking of t_m at $P =$
0.99 and the degree of freedom $(n - 1)$ shows that the values do not exceed
2.807. It confirms the validity of the calculations performed and the fact
that the differences received are not significant.

Photoelectric, optical, and electromechanical transducers are used for
control and measurement of the strokes of machines and instruments and
can be taken as an example of the information on displacement genera-
tors. However, they experience significant systematic errors when being
applied in these machine systems with long displacement strokes. A technical solution incorporating the aforementioned advantage of lessening
the indeterminacy of the measurement is through the use of multiple indicator heads with a short part of the scale which can be calibrated at very
small steps with high accuracy. The indicator heads of a measurement
system must be placed in the path of a moving machine part in such a way

that the measurement be limited only to the most precise part of the measuring system (scale), and the reading of the information be performed serially by using the heads placed along this path.

As an advantage of such a system, it is possible to measure a short length of the scale with higher accuracy and with a smaller pitch of measurement. So, in such a case all the length of this part of the linear translational transducer is metrologically determined, that is every stroke or the full length of the measuring part of the transducer is determined with appropriate uncertainty. The example discussed above shows a systematic error correction by technical and informational means. The uncertainty estimate, as random constituent, cannot be so corrected. It must include contributions from all sources, including primary sampling, sample preparation, and chemical analysis, in addition to contributions from systematic errors, such as sampling bias. Reliable estimates of the uncertainty around the concentration values are used in chemical, medical, and biological measurements. So, it can be stated that such attributes as area, mass, concentration, volume, humidity, and so on, in many occasions have an influence in sampling and measuring result evaluation. Hence these attributes would be useful to include in the final result, or, at least, to show their influence in the measuring procedure.

The same situation occurs in technical measurements, especially in determining a measurand with multivalued results, as with CMMs, GPS measuring systems and GPS receivers, encoders used in robotics and CNC machines, and so on. It is evident that the sampling problem is important and it is worth further development. It is a basis for data processing in all measurements and it is widely used in all kinds and branches of metrology. So, it is important to determine the information quantity on an object that was assessed providing more complete measurement during the measurement processes preparation.

The investigations discussed above prove that a new approach to the sampling strategy and the information quantity derived from the measurement process must be implemented for scientific and practical purposes. Additional information of the sampling value for the result of measurement would expand the knowledge of the extent of evaluation of the object. It would also help for data comparison and for achieving more exact assurance of traceability. We express here an idea to include the sampling procedure into the equation for the measurand and that would show a full procedure of the measurement process.

Besides the standards for the measurement data uncertainty evaluation, assessment and processing, there are standards for the selection of samples for testing and measurement. The guide to the expression of

Uncertainty in Measurement [13], usually referred to as the GUM, is published by ISO and remains the main document for uncertainty evaluation. Documents concerning sampling procedures are directed for preparation and performance of testing, and there are no indications transferred into the result of measurement and how this process was carried out.

It is indicated that a significant discrepancy can occur between the analytical conditions of a routine measurement and the analytical conditions that were used in clinical studies upon which the decision-making criteria are based. This can lead to serious interpretation errors with relevant consequences. The correlation of used methods with the reference method is described showing also some overestimated and underestimated results.

Bearing in mind the many varieties of information assessment during a measurement, it is evident that the sampling problem is important and is worthwhile to develop. The most common example used in all fields of metrology is calibration of weights. The scale of electronic weights cannot be calibrated at every 0.1 mg or so. A calibration process is carried out at some pitch of the scale range. So, some intervals between the calibrated points on the scale are left undetermined. It is an unavoidable problem without solution as there are no means for calibration at every stroke of the scale or every digit of the information-measuring system of the weights.

During the calibration of scales it is possible to determine only a restricted number of values from the full range of existing data. Also suitable sampling of measuring points on the surface of industrial parts was shown to be a very important task, as was the sampling of the points in a machine's working volume during its calibration. The accuracy of the calibration must remain at the same level, but the time consumed must be minimized.

Technical problems related to sampling in measurements are discussed in many research papers, and sampling procedures are important in dimensional measurements, measurement of geometrical features, and position between them. The main approach in this procedure is probability theory which provides a statistical means for evaluating the results of measurement (the measurand) by selecting the pitch of measurement that serves as some kind of sampling, assessing the set of trials, calculating the mean value of estimates, and evaluating the dispersion at the probability level chosen. It is the basis for data processing in all measurements and is widely used in all branches of metrology.

A general diagram of the process is given in Figure 1.12. The reference material has attributes $Q_j = f_j (n_k, v_l, \zeta_p, ...)$ that depend on, for example, a number of samples of mass, volume, humidity, and so on. The

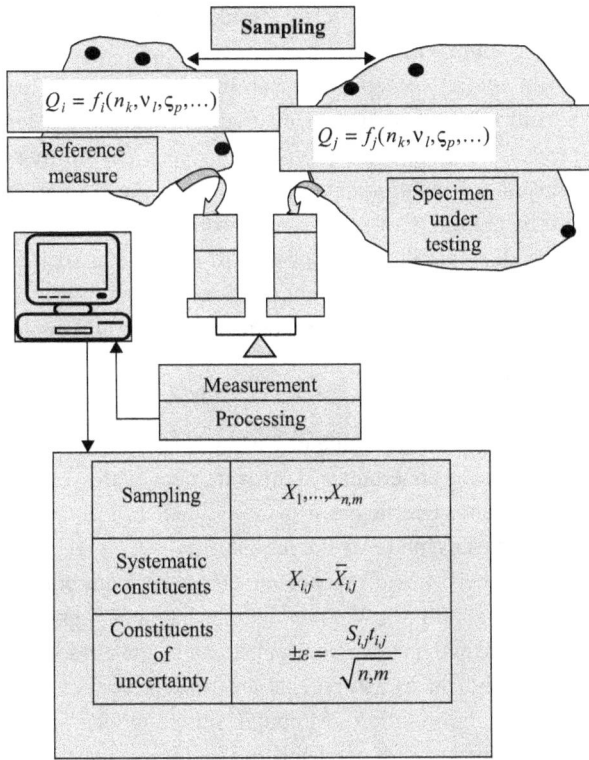

Figure 1.12. General approach to measurement and the result evaluation.

material under testing has the same kind of attributes $Q_i = f_i(n_k, v_l, \varsigma_p, \ldots)$ depending on different quantities of the same attributes. The strict statistical evaluation of these materials for comparison will be true, but it can have some discrepancies depending on those attributes that exist at the point of evaluation. The process presented here would give the information on the conditions of measurement performed, as it will contain the attributes of sampling, systematic error, and uncertainty evaluation.

The general approach to the enlargement of information processed during the measurement process is presented as follows. At first the sampling procedure $X_1, \ldots, X_{n,m}$ is undertaken. Here X—variable, n, m—number of trials for sample estimate. The next step will be determination of the systematic constituents $X_{i,j} - \overline{X}_{i,j}$ both for the reference measure i and the measure under testing j. In the case of a reference measure that is already determined, it is performed only for the testing object. By a similar operation the uncertainty is determined $\pm \varepsilon = S_{i,j}\, t_{i,j}/\sqrt{n, m}$ where ε—uncertainty,

$S_{i,j}$—estimated standard deviation, $t_{i,j}$—Student's coefficient, n—number of attribute i, m—number of attribute j.

There is a widely accepted probability level of 1σ, 2σ, and 3σ (0.68, 0.95, and 0.99) in statistical evaluation processes, where σ represents the true standard deviation. With many technical information measuring systems an information determination level equal to 50 percent of total data can be chosen. However, some information systems, consisting of gigabytes of data, would be impossible to evaluate at such a level. In those areas it would be possible to choose 5 to 10 percent of data evaluation. The expression of the measurement result is widely accepted as

$$X = \bar{x} \pm \varepsilon, P,$$

where X is the result of measurement; $\bar{x}$ is the systematic part of the measurement result, ε is the uncertainty of measurement expressed as $\varepsilon = ts/\sqrt{n}$, where t is the Student's coefficient, S is the estimate of standard deviation, n is the number of trials, and P is the probability.

There are three vertical branches covering the main constituents of data assessment evaluation in the concluding diagram, Figure 1.13. Two main branches contain the constituents, systematic and random, according to which most technical measurements are presented. The third branch includes sampling information. Although sampling characteristics are used more widely in chemical, medical, and biological measurements, the same efforts have recently been made in the field of technical measuring systems.

An important task is to evaluate the information required for measuring multivalued objects by assessing systematic and random errors as usual, and including the information entropy of measurement, that is indicating the quantity of measurement information that was evaluated from the total accessible information. Although many authors have discussed the task of uncertainty monitoring and sampling, there is no general approach given to solve this task. Some ideas have been expressed in this work to include the sampling procedure into the equation for the measurand. This is especially important bearing in mind the different chemical and biological, composite, nonhomogenous materials, and so on. Also modern information measuring systems consist of smart transducers that can combine a wide range of data or measurement values reaching tens of thousands of numerical values (e.g., from a laser interferometer). This process of including sampling information into the equation for the measurand is also important in view of traceability of the measurement as it can clearly indicate which part of the information was assessed during the sampling and measurement operations.

RESULT OF DATA EVALUATION

SYSTEMATIC CONSTITUENT

RANDOM CONSTITUENT (UNCERTAINTY)

SAMPLING INFORMATION

$y = (ax + b)$,

Measurement uncertainty

NO INFORMATION

$y = A + cos(\omega t + \varphi)$

Uncertainty due to environment influence ε, P

L-P sequences gray theory

Other periodical function

Uncertainty of the standard measure ...

Hammersley distribution spline approximation W-shape sampling

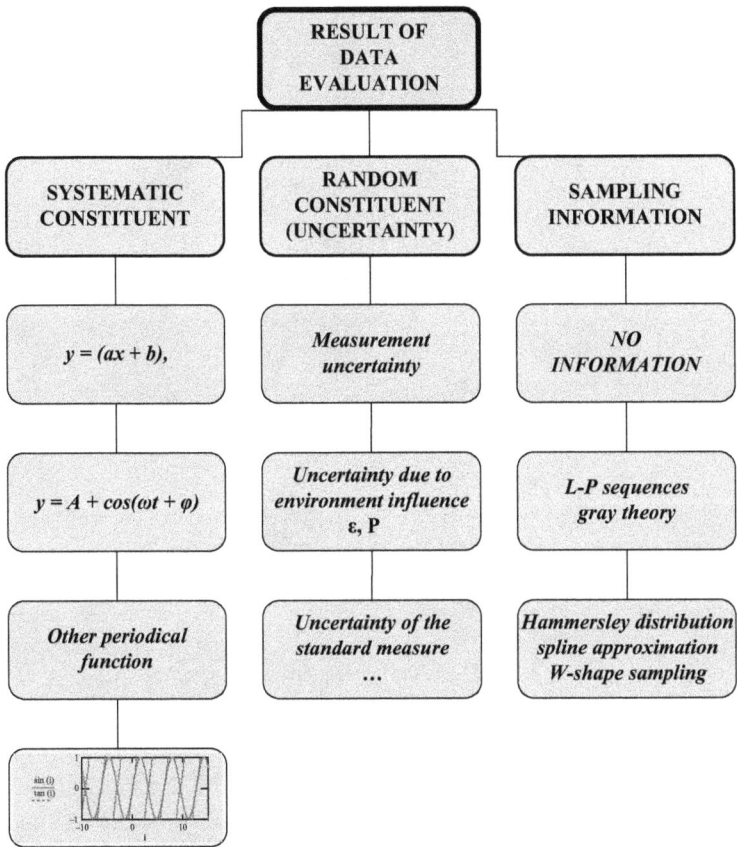

Figure 1.13. The generalized diagram of presenting data evaluation.

It is evident from the investigations mentioned above that a new approach to sampling strategy must be investigated and implemented. The problem remains in the field of development of calibration methods that allow us to obtain a large quantity of the accuracy data in a short period of calibration and using simple reference equipment of low cost. The second step in this field would be the efficient use of data selected for the accuracy improvement of machines.

CHAPTER 2

TESTING OF GEOMETRICAL ACCURACY PARAMETERS OF MACHINES

The geometric accuracy parameters of machines and instruments are discussed in this chapter, and various parameters, such as straightness, perpendicularity, flatness, pitch, yaw, roll, and so on, are introduced. The 21 detailed geometric errors of three coordinate machines together with some supplementary parameters are covered in detail. The principal processes of measurement of these parameters are explained and arrangements for their control are discussed.

Machine accuracy testing methods common for many computer numerically controlled (CNC) machines, coordinate measuring machines (CMMs), and measuring instruments are covered. CMM tests, as specified in ISO standards, include acceptance tests, reverification tests, interim tests, and full parametric calibration tests. Testing of the accuracy parameters of rotary tables is also discussed at the end of the chapter.

2.1 GEOMETRICAL ACCURACY PARAMETERS OF MACHINES AND INSTRUMENTS

The most advanced methods are used for the assessment of the accuracy parameters of machines and measuring equipment, such as the CMM—due to its high-accuracy parameters and its use for quality control in research and industrial production. Different optical methods applied for three-dimensional (3D) measurements are investigated including the Projected Fringes Method, Electronic Speckle Interferometry, Structured—Lighting Reflection Techniques, White Light Interferometry, and Laser Scanning methods. Instruments for fast 3D detection of surface points and numerical presentation are continuously under development, and the advantages of fiber optics, fiber sensors, and measurement systems based on these are used. The use of optical means for displacement and position measurements gives great advantages due to its minimal dimensions, wide range of accessibility, integration of the sensor and information transfer functions in one unit, and so on.

It is noted that these modern techniques are not used widely enough in CMM calibration and accuracy verification processes of CNC machines. Special reference parts (artifacts) are used for CMM calibration purposes with their entire dimensions assessed beforehand with high accuracy. Many investigations are made to determine the measuring coordinate system uncertainty using datum planes, cylindrical and spherical bodies, and sets of discs placed and calibrated in the measuring volume of the machine. A variety of configurations of such artifact parts are used. They include rectangular plane surfaces, internal cavities or exterior spherical bodies, cylindrical and conic surfaces, and so on. The disadvantages of such methods of accuracy control are in the restrictions of the measurement volume of the machine to be assessed. Most of such reference parts are of small dimensions, so the calibration does not cover all the volume of the machine. It is inconvenient to handle a reference part of large dimensions on the table of the machine. In this case, a large volume inside the machine is unavailable for calibration. Consequently, some new methods and means for easier and more convenient calibration of the accuracy parameters should be investigated, developed, and introduced into practice.

A general review on the geometrical accuracy parameters of machines can be accomplished using the symbols accepted for their denomination, as shown in Table 2.1.

All these parameters are multiplied by splitting them into the coordinate axis of the machine, so making the checking more complicated. For the accuracy verification of CMMs, it is important to choose the etalon or reference measure for each of the parameters, for example, a reference length measuring system as the standard measure to calibrate all three coordinate axes of a CMM. The precision length measuring system is chosen from a wide variety of reference measures used for this purpose. From the variety of length standards, a laser interferometer is used most often as a reference measure. Together with high accuracy of the reference measure, there are other requirements to be followed during the accuracy verification of CNC machines, one of them being the importance of avoiding the influence of Abbé offset, cosine error, and tilting errors occurring during the measurements. It is important to determine errors which act in

Table 2.1. Symbols for indication of geometrical characteristics of machines

Characteristics	Symbol	Characteristics	Symbol
Straightness	—	Flatness	▱
Roundness	○	Cylindricity	⌀
Profile any surfaces	⌓	Profile any line	⌒
Perpendicularity	⊥	Parallelism	//
Position	⊗	Symmetry	÷
Concentricity coaxiality	◎	Angularity	∠
Total run-out	⌰	Circular run-out	↗

the different directions of coordinate axes in the measuring volume. For many of the machines the measurements are performed not only along the directions of coordinate axes, but also in the diagonal direction of the measuring or working volume of the machine. The diagonal method is efficient because of involving three dimensions in every step of the measurement performed. There are some methods of calibration where the redundancy technique is employed. The calibration in the area or volume is performed placing the reference measure at several different positions. An optical scale can be used for such purpose. The systematic errors of the machine are identified and after mathematical analysis the errors are then corrected by some technical means.

The challenge that remains in this field is to develop calibration methods that capture a large quantity of the accuracy data in a short period of calibration using simple low-cost reference equipment. The second challenge in this field would be the efficient use of data selected for the accuracy improvement of the machines. These items and some methods and means for this purpose are to be discussed and applied in monitoring operations.

Traceability and calibration of the accuracy parameters of the parts of multicoordinate machines, robots, and CMMs is quite a complicated task to accomplish and assess. The large amount of information in 3D measurements and the lack of a spatial reference measure are the main technical and metrological tasks to overcome. Some technical methods for the measurement of accuracy elements are presented here, and some nontraditional methods and techniques for complex accuracy assessment are proposed.

The accuracy of geometric parameters of precision machines, such as metal cutting tools and CMMs, is quite a complicated task to assess because of the variety of accuracy parameters to be checked and the high accuracy that must be assured. It is a time-consuming process that requires very experienced and well-qualified staff to perform the calibration and several reference measures to achieve the traceability of the calibration. CNC machines have some additional features, which add to the complexity of overall calibration problems, such as the accuracy parameters of the rotary table, position of the cutting instrument, and instrument changing devices. For CMMs there are also the accuracy parameters of the measuring head (touch trigger probe). The list and tables of geometrical accuracy parameters together with diagrams of arrangements for their control and accuracy verification are discussed.

The position of the moving parts of CNC metal cutting tools and technological machines are determined by geometric accuracy parameters in respect of six degrees of freedom (DOF). These accuracy parameters consist of a separate group of technical parameters of the machines that need separate and sometimes rather special means of measurement for testing or monitoring. The six DOF are often called generalized coordinates. These parameters are described by technical specifications on the machine and special written standards for separate parameters or for total accuracy verification. A set of instruments is used with different classes of accuracy (higher, of course, than the accuracy of the geometrical parameter to be checked), having special ranges of measurement, sensitivity, resolution capability, and so on. Modern machines have many coordinate movements, much more than six of them, so the number of DOF will also be higher. Some parameters are too complicated for measurement and sometimes there are no means for measurement of one single parameter without the influence of the others. An example is the measurement of straight-line movement of a part of the machine by the use of a reference measure of the straight line in the form of the flat surface of a reference measure made from steel or granite. A contact inductive gauge usually is used for the measurement of this parameter, and during this measurement the other parameters, such as the pitch of movement of the moving part can also influence the results of measurement. In the case, when more exact accuracy analysis is needed or when there is a need to investigate separate accuracy parameters, then additional measurements are required using different methods and means. The measurement mentioned above must be repeated using an autocollimator and reflecting mirror, which is not suitable to measure the displacements perpendicular to the movement direction.

The grinding machine tool shown in Figure 2.1 is an example of a precise machine with accuracy parameters that must be measured or monitored. It is designed for grinding the inner and outer cylindrical surfaces, tapestry, end surfaces, and step cylinders. The accuracy of geometrical parameters composes the datum for forming the high-accuracy coordinate displacements of the machine's components and smoothness of surfaces of the workpieces manufactured. The datum of the machine transfers its geometrical features to the part produced.

Some of the technical and accuracy parameters of the machine are shown below:

- Maximal length of the part for grinding, 180 mm;
- Maximal outer diameter of the part, 100 mm;
- Maximal diameter of the hole for grinding, 40 mm;
- Value of the single feed step, 0.002 mm;
- Value of the periodical pulse feed, 0.001...0.01 mm;
- Angle of rotation of the grinding head, ± 30 degrees of arc;
- Maximum angle of rotation of upper table, ± 6–7 degree of arc.

Figure 2.1. The cylindrical grinding machine of high-accuracy model 3U10MAF10 (J/V VINGRIAI).

These and other technical specifications assure very high accuracy of geometry and high smoothness of the surfaces of the workpieces machined. For example, run-out of the cylindrical surface can be achieved within the range of 0.2 to 0.3 μm. The control unit of the machine transfers an accuracy of the linear displacement transducer (LDT) feeding the grinding wheel to the required value that can be controlled in two coordinate directions. Widely used LDTs for CNC machines are photoelectric raster scales with a reading head which after an interpolation assure an in feed value of 0.001 mm (or rather 1 μm). Geometrical accuracy parameters also include the high geometric accuracy of slide ways, their mutual parallelism or perpendicularity, front and back grinding heads' coaxiality, and so on. All these parameters are to be checked under the production output from the plant and later on during service in industry. This is determined in the written standards on the relevant metal cutting tools and also by technical specifications on the machine. Many accuracy parameters are similar for various types of metal cutting tools, such as milling, grinding, coordinate drilling, and grinding; and for some measuring equipment, such as three-coordinate instrumental microscopes, roundness measuring instruments, CMMs, and others.

The total accuracy parameters common for many metal-cutting machines and instruments, and which must be measured or monitored during their performance verification, are presented below. Two main groups of these parameters are normalized in the written standards. They can be selected in groups by detailed parameters and complex accuracy indices.

Such parameters can be treated as a group of elements:

- Straight-line movement trajectories in all coordinate axis and in the two mutual perpendicular planes along these axis, total six parameters;
- Positioning accuracy by moving in the three coordinate axes with period stops at the predetermined positions, three parameters in the three perpendicular (Cartesian) axes;
- Mutual perpendicularity of trajectory of moving machine's parts, three parameters;
- Angular stability of the machine's part by movement along the three coordinate axes in the three perpendicular planes, nine parameters.

So, a total of 21 parameters (see Figure 2.2) are normalized for the control of the machine's accuracy. Also, dependent on the specific features and a number of coordinate axes (DOF) in the machine, there can be some additional parameters normalized and checked or monitored. These can

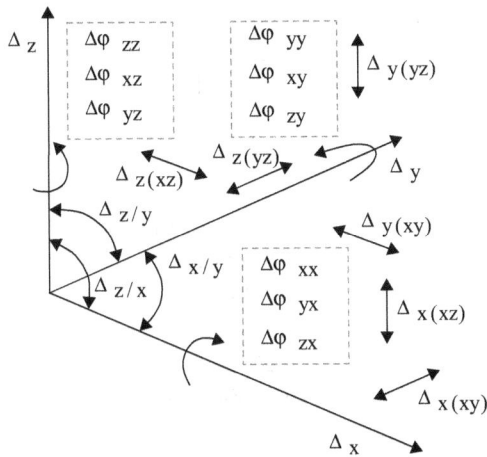

Figure 2.2. Distribution of geometrical accuracy deviations in the coordinate axes of the machine.

include axial or radial run-out of the axis of the rotary table, if present; accuracy parameters of the spindle, if there is a rotating spindle in the machine; additional angular movements of the robot's joints in the variety of displacements in space, and so on. Some industrial robots, such as painting and assembly robots, have many DOF, and subsequently the number of accuracy parameters grows in an appropriate manner. In this chapter we are mostly concerned with the analysis and testing of CMMs, as they are typified by their three coordinate movements, information/measuring systems for the control of these coordinate movements, sensors/measuring heads or touch-probes, and so on.

The moment when the instrument of the machine (drill, cutter, tip of the touch-probe) touches the surface of the workpiece must coincide with the position of coordinates preset into the information—measuring system. Any deviation from that position means an error of performance of the machine.

During the complex accuracy checking, the workpiece manufactured on the machine is measured by an equipment of higher accuracy, for example, by a CMM. Sometimes a special workpiece, or artifact, is used as a representative of all technical parameters of the machine to be checked or monitored. An artifact is a useful reference tool for fast and efficient monitoring of the parameters of the machine to be controlled. In fact, the main coordinate displacements are designed to be included into the form of the artifact.

Two main steps are used for the accuracy control of multicoordinate machines, robots, and measuring equipment. The first one is to calibrate the accuracy of geometric elements of the machine, that is the straightness, squareness, flatness of guide ways, and bedding; the bias of vertical, horizontal, and longitudinal traverse, pitch, roll, yaw, and so on. The second step consists of measuring the reference part calibrated beforehand. The results of the measurement of such a part on the CMM are compared with the accuracy results of this part collected during its calibration. Geometric features are assessed while aligning the coordinate system during the calibration by the producer and later by the customer's quality control staff. The alignment of the machine's coordinate system and the correction of the systematic errors are mostly dependent on geometric errors and their distribution characteristics.

The errors listed above must be controlled by the methods specified in the published standards or according to the standards or technical specifications of the producer. Special accuracy requirements are for the measuring head of the CMM or position of working instrument of the machine. The great number of accuracy parameters, methods, and instruments makes the task of calibrating multicoordinate machines very difficult, complicated, and expensive. Geometric features are assessed during coordinate system alignment during the calibration by the producer and later by the customer or quality-control staff. There are many investigations made to determine measuring coordinate system uncertainty using datum planes, cylindrical and spherical bodies placed and calibrated in the measuring volume of the machine. The alignment of a machine's coordinate system is mostly dependent on geometric errors and their distribution characteristics.

The geometric errors in the volume of a multicoordinate machine consist of the perpendicularity of coordinate axes $\Delta_{x/y}$, $\Delta_{x/z}$, $\Delta_{y/z}$ (Figure 2.2), the coordinate position errors $\Delta_{x,y,z}$ along the axes x, y, and z; rolling errors $\Delta\varphi_{x,y,z}$ around the axes x, y and z, pitch and yaw errors $\Delta_{x(y,z)}$ during the movement of the part along the relevant axis in the indicated plane, and so on. So, there are 21 detailed geometric errors of three-coordinate machines. The other errors on the machine, due to environmental influences (temperature, pressure, dynamics, etc.), are analyzed and assessed in other investigations. Specific errors exist when a rotary table is present on the machine and due to the signal formation of the measuring head of a CMM, and they are analyzed later on in this chapter.

The 21 detailed geometric errors of three-coordinate machines together with some supplementary parameters are listed in Table 2.2.

Table 2.2. The parameters of geometrical accuracy of the moving parts of the machines

No	Symbol	Accuracy parameter
1.1	Δ_x	Coordinate displacement during periodic stops (positioning), deviation in axis x
1.2	Δ_y	Coordinate displacement during periodic stops (positioning), deviation in axis y
1.3	Δ_z	Coordinate displacement during periodic stops (positioning), deviation in axis z
2.1	$\Delta x(xy)$	Deviation from straightness of motion of x axis in plane xy
2.2	$\Delta x(xz)$	Deviation from straightness of motion of x axis in plane xz
2.3	$\Delta y(xy)$	Deviation from straightness of motion of y axis in plane xy
2.4	$\Delta y(yz)$	Deviation from straightness of motion of y axis in plane yz
2.5	$\Delta z(xz)$	Deviation from straightness of motion of z axis in plane xz
2.6	$\Delta z(yz)$	Deviation from straightness of motion of z axis in plane yz
3.1	$\Delta_{x/y}$	Deviation from perpendicularity of motion of x axis and motion of y axis
3.2	$\Delta_{x/z}$	Deviation from perpendicularity of motion of x axis and motion of z axis
3.3	$\Delta_{y/z}$	Deviation from perpendicularity of motion of y axis and motion of z axis
4.1	φ_{xx}	Roll deviation during the displacement along coordinate x
4.2	φ_{xy}	Yaw deviation during the displacement along coordinate x in respect to coordinate y
4.3	φ_{xz}	Pitch deviation during the displacement along coordinate x in respect to coordinate z
4.4	φ_{yy}	Roll deviation during the displacement along coordinate y
4.5	φ_{yx}	Yaw deviation during the displacement along coordinate y in respect to coordinate x

Continued

Table 2.2. (*Continued*)

No	Symbol	Accuracy parameter
4.6	φ_{yz}	Pitch deviation during the displacement along coordinate y in respect to coordinate z
4.7	φ_{zz}	Roll deviation during the displacement along coordinate z
4.8	φ_{zx}	Yaw deviation during the displacement along coordinate z in respect to coordinate x
4.9	φ_{zy}	Pitch deviation during the displacement along coordinate z in respect to coordinate y
Supplementary parameters		
5	Δ_{gi}	Position deviations of the instrument (measuring head or touch-probe in case of measuring machine) during the touch on the surface
6	φ_{l}	Axis run-out of the rotary table
7	$\Delta_{il,s}$	Mutual parallelism or perpendicularity of the moving parts of the machine
8	Δ_{s}	Axial or radial run-out of the spindle
9		Other supplementary parameters, such as cylindricity, mutual parallelism or perpendicularity of axes; parallelism or perpendicularity of axes to the datum or some trajectories of displacement, etc.
10	Δ_{ki}	Complex testing deviations

The tests are performed mainly during the machine acceptance process, calibration, and periodic inspections. Monitoring of the accuracy parameters can be accomplished mainly by using the relevant artifacts, such as flatness standards, step gauges, ball plates, and space frames, which represent the main accuracy parameters or their variations (for example, perpendicularity, flatness, straightness, length and angle standards, or other sets of parameters).

2.2 ACCURACY CONTROL OF GEOMETRICAL PARAMETERS

EN ISO, BS, and ASME written standards [14–16] provide definitions and specifications on machine accuracy testing methods and means for their performance. This chapter will only cover those which are common for many CNC machines, CMMs, and measuring instruments.

The preparation for testing of the accuracy parameters is accomplished by setting the standard measure along the line of measurement and fixing the gauge interacting with the standard at a relevant position. All means of measurement must be aligned along the measurement axis and they must be set at a minimal height from the measuring surface. This requirement was formulated by physics scientist *Ernst Abbé* (1840–1905) who introduced the concepts of Abbé error and Abbé offset (Figure 2.3). Abbé offset is a misalignment between the axis of displacement of the moving part of the machine and the axis of the standard of measure. This is valid for the testing of the geometrical parameters of the machine when the standard measure is placed at some distance from the axis of motion. The Abbé error is an error that occurs because of angle deviations of a movable component at the length of Abbé offset between the machine's measuring system and the measurement line. Abbé offset is valid for every displacement at every position in the space, along every coordinate. Figure 2.3 shows an Abbé error and offset when the movable component of the machine reads the scale of its information-measuring system whose axis doesn't coincide with the axis of the scale placed for testing the positioning accuracy. The same phenomena also occur due to the angular shift of a measuring instrument or gauge, especially when it is fixed on a lever having significant length.

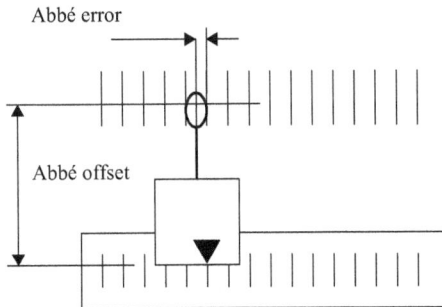

Figure 2.3. Abbé offset and error during measurement in the test line direction.

Some further definitions (EN ISO 10360-2:1995) are presented below which are also important for most general measurements as well as for machine performance monitoring [14].

1. **Direction of the test line:** This is a direction of every line which is parallel to the line of movement of part of the machine. It can

be determined by least square method calculations assessing the deviations due to errors of perpendicularity, parallelism, angular displacements, and so on.

2. **Cosine error:** This is the measurement error in the motion direction caused by angular misalignment between the axis of displacement and the axis of the standard measure to which a relevant parameter is compared.

3. **Measurement error of the length** E**:** This is expressed in micrometers and presented in one of these forms:
   a) $E = $ minimum of $(A + L/K)$ and B, or
   b) $E = (A + L/K)$, or
   c) $E = B$,

   where A is a positive constant, expressed in micrometers and supplied by the manufacturer;

   K is a dimensionless positive constant supplied by the manufacturer;
   L is measured length in mm;
   B is the maximum value of E in micrometers supplied by the manufacturer.

   Measurements are performed by using all three coordinates of the machine according to the manufacturer's statement on applying the E value.

4. **Error of indication of a CMM:** This is a value of indication of a CMM display unit minus the (conventional) true value of the measurand. Angular deviations along the axis of movement are shown in Figure 2.4.

5. **Error of indication of a CMM for periodic reverification:** This is an error, chosen by the user for the periodic reverification test according to the user's requirements and use of the machine.

6. **Periodic error:** An error that is periodic over an interval, usually, under an influence of periodic errors of the measuring system, drive, bearings, or other components of the machine.

7. **Perpendicularity:** This is the deviation of one linear motion against a geometrical straight line perpendicular to a geometrical straight line, which is the reference for straightness of motion.

8. **Probe (measuring head, touch-probe):** A device which establishes the coordinate position of the movable components of a machine relative to a measurement point.

9. **Straightness:** This is the deviation from a geometric straight line in the linear motion of each axis.

10. **Probing error,** R**:** This is the error, within which the range of radii of a material standard can be determined with a CMM, the measurements being taken using a sphere as a substrate. The value

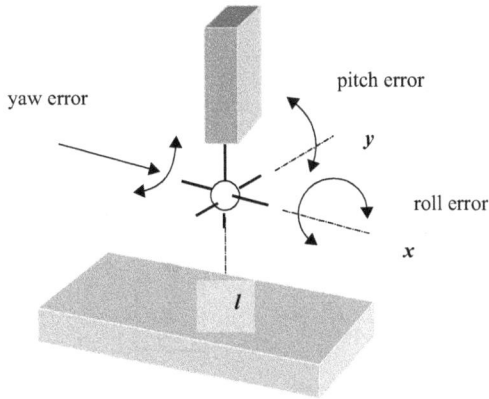

Figure 2.4. Angular deviations along the axis of movement.

of **R** applies for any location of the material standard in the measuring volume of a CMM.

11. **Pitch deviation (Figure 2.4)**: Angular deviations along the moving axis in the vertical plane.

12. **Roll deviation (Figure 2.4)**: Angular deviations along the moving axis in the plane that is perpendicular to the moving axis.

13. **Yaw deviation (Figure 2.4)**: Angular deviations along the moving axis in the horizontal plane.

14. **Repeatability:** The ability of an instrument to produce the same indication when sequentially sensing the same position under similar measurement conditions.

Measurements of geometrical accuracy parameters are performed by using a relevant reference measure (etalon) and mechanical, electrical, or opto-electronic gauges. One of these instruments (gauge or etalon) is fixed and adjusted to the part of the machine to be tested, the other one is fixed to the constant support (bedding of the machine, the stand fixed on the ground). The sensor part of the measuring gauge is under the interaction (contact or noncontact) with the etalon. Etalons or reference measures can be

- mechanical—metal, granite (or of other material) straightedge or template, straight-line rectangles, cubes, polygons, triangles; measuring rules, end or angle gauge blocks, artifacts;
- optical, opto-electric, laser interferometers, autocollimators with reflecting mirrors or polygons, photoelectric microscopes;
- electromechanical, inductive, capacitance gauges, or piezoelectric transducers; accepted as reference measures for special kinds of measurement.

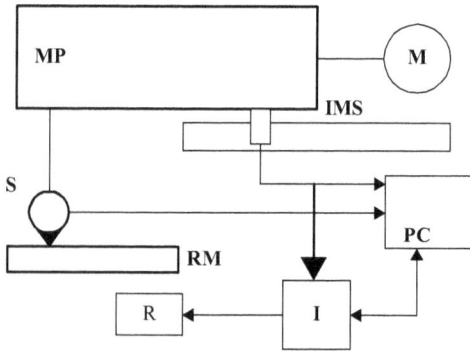

Figure 2.5. Generalized diagram of measurement (monitoring) of geometrical accuracy parameters of the machine.

A generalized diagram of measurement of geometrical parameters of machines is shown in Figure 2.5. The system consists of a moving part of the machine (MP), of which movements (trajectories, positioning, etc.) are to be measured, the information/measuring system of the machine (IMS), one part of which is fixed to the constant base, the other part to the MP, a reference measure (RM), used as an etalon for comparison with the measured parameter, sensor S, usually fixed to the object to transfer the difference between the reference measure position or values and the object's real values, a PC or other control unit to control the process, registration unit R and often supplementary units I for interfacing information transferred from the sensors and other measuring or monitoring units to the control unit. The devices used for the measurement and monitoring of geometrical accuracy parameters should have their accuracy three to five times higher than the range of errors for measurement on the object. Also a displacement, position, and adjustment in all directions in the space must be appropriate for the tasks of measurement to be fulfilled.

Machine performance tests are held at specified set periods or under special requirements, for example, after evident failures, accidents, functional failures, or loss of accuracy. A predetermined number of procedures are foreseen for this purpose. The parameters to be tested can be illustrated by CMMs or grinding machines, as they require the highest accuracy parameters and feature a wide variety of DOF used in their design.

EN ISO standards specify the following CMM tests:

- Acceptance tests, during which the performance parameters of the machine and measuring system are required to comply with the technical specifications declared by the supplier.
- Reverification tests (periodic inspection), during which the user has an opportunity to check periodically a machine's performance parameters and the measuring system.
- Interim tests during which the user can check the performance of the CMM and its probing system between regular reverification tests.
- Calibration, performed as parametric calibration.

Accuracy tests are included into all kinds of tests as accuracy parameters have the most influence on a machine's performance ability. They are made using the length end blocks and gauges, metal linear scales with microscopes and laser interferometers. The tests are performed in various positions in the working volume of the machine.

Before the beginning of tests the CMM must be put into operation according to the producer's recommendations, the measuring head is identified and its compatibility checked. The working and environmental conditions must also be set under the technical specifications on the machine. Minimal requirements can be fulfilled using some machine monitoring tests. The use of artifacts for machine verification can be considered as one of the possibilities for the machine monitoring processes. The main task of these procedures is to keep equipment at the required level of accuracy and at its proper level of performance. The tests are performed under a mutual agreement between the producer and user of the machine.

Basic machine tests are used to check how the length measurements are performed on the machine. The error E is checked by measurement of the length gauges.

2.2.1   LENGTH (POSITIONING) MEASUREMENT

This is one of the accuracy tests within CMM performance verification. It is performed by measurement of the accuracy of linear displacement in three coordinate axes using the material standards of length as gauge blocks and step gauges, a reference linear scale or a laser interferometer. The measurement is for checking the traceability of CMM length measurements to the international standards of length. Different length measures are recommended for this test. It is recommended that

- the longest length of the material standard is at least 66 percent of the longest diagonal of the working volume of the machine;
- the shortest length of the material standard is less than 30 mm.

If the user's material standard is used, the length measurement error must be not greater than 20 percent of the value E. If it is greater, then E must be redefined as the sum of E and this uncertainty. In the case of use of the manufacturer's material standard, then no additional uncertainty shall be added to the length measurement value E.

Measurements are performed in every direction of the working volume of the CMM at the user's discretion, making bidirectional measurements either externally or internally. Five test lengths are chosen, each measured three times. The error of length measurement is calculated as the absolute value of the difference between the indicated value of the CMM and the true value of the relevant test length. Gauge blocks (Figure 2.6) are widely used as material length standards for this purpose. Supplementary measurements are usually made during the alignment process.

Measurements for position accuracy tests are taken along one measurement line for each axis. The reference length measuring system, gauge blocks or laser interferometer, is aligned to each machine axis within a permissible deviation of 1 min of arc. The reference measure is situated at the center of measuring travel according to all coordinate axes.

The gauge blocks used for measurement must not be deformed, the direction of their placement must coincide with the axis of travel, and a cosine error must not exceed 10 percent from the linear displacement tolerance of the machine.

Measurement intervals must not exceed 25 mm or less when the measurement travel is 250 mm. When the measurement travel is about 1000 mm the measurement interval must be at least 25 mm, but not exceeding

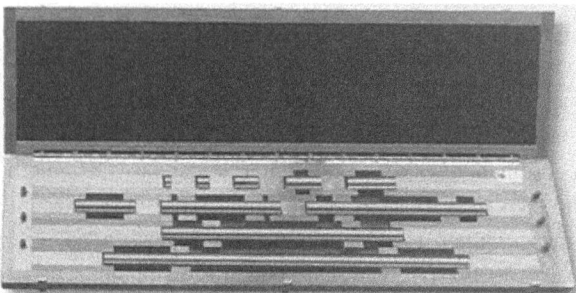

Figure 2.6. Set of length gauge blocks.

1/10 of travel length. For the travel length more than 1000 mm, the measurement interval must not exceed 100 mm. Accuracy specifications of the CMM must be noted at such intervals and at least 20 positions would be measured. During the measurements the CMM measuring head that will be used for most measurements is used. The readings of the CMM must be reset to zero at the first touch to the surface of the length gauge. Three measurements must be performed at every measurement interval within the travel length and the mean arithmetic value is calculated from the results obtained. This is repeated along every coordinate axis. The environmental influences must be minimized by trying to perform the measurements in the shortest time possible.

Measurements using a laser interferometer must be performed in an environment free of shock and vibrations, at constant temperature. The reflector of the interferometer is fixed to the movable part of the machine or component and aligned according to the laser's light beam. Usually the ambient and material's temperature, humidity, and pressure are measured by supplementary gauges which provide input into the control unit of the interferometer. Correctional coefficients are calculated and the final result of measurement is presented taking account of these corrections. The length or positioning measurements can be supplemented by laser diagonal length measurements. Diagonal measurements are taken along any space diagonal of the measuring volume. The result of measurement is also assessed as the difference between the distance readout of the machine and the value of the laser measuring system. The error of a CMM includes the error of the material standard of length and is assessed including an uncertainty of measurement that characterizes the range of values within which the true value of a measurand lies. Measurements in one, two, and three-dimensional directions are applied for the uncertainty determination, the difference being in the displacement in 1D, 2D, and 3D coordinate directions. Relevant coordinate measuring systems of the machine are present during such measurements.

The verification of the machine is approved when all but one single measurement from 105 does not exceed the value E indicated by the manufacturer and expressed in micrometers. At least 5 from 35 trial measurements of length can exceed the value E. Every such measurement which exceeds the value of tolerance must then be repeated 10 times.

Repeatability measurements should involve assessment of all systems of the machine including the mode of operation, work of the operator, and software performance. The test conditions should be as close to the working conditions as possible. This is why there are requirements that different tests should be undertaken using different modes of operation. A modified test procedure must be applied for a machine with a large working volume. In special cases when they do not fulfill the requirements of the standard, then the supplier and user shall agree the selection of tests.

Repeatability tests can be executed using a precision reference spherical body mounted on the machine's table, approximately at the mid-point of the machine's travel axes. The measurements are performed by determining the ball center coordinates. At least 10 sets of four contacts to the ball surface should be made and the center coordinates of the ball calculated. Repeatability is determined as the maximum spread of the measured center values in all three coordinates. Repeatability tests can be performed under computer control mode or manually.

The probing system testing consists of using the probe (measuring head) of the CMM to be tested and a material reference spherical body as a substrate. The probe configuration and the tips are selected under measurement requirements. The diameter of the spherical body is between 10 and 50 mm. The most used diameter is 30 mm, and the calibrated error of its total run-out must not exceed the value R (R—probing error). It is recommended to take at least 25 random point measurements on the surface of the reference sphere approximately evenly distributed on the surface. Applying the least squares method, the center of the sphere is calculated and it is compared to the value stated by the manufacturer.

The main methods and means of measurement of the accuracy of geometrical parameters of machines and CMMs are presented in Tables 2.3 and 2.4.

Some other errors due to the influence of environment temperature, pressure, dynamics, and so on, act outside the machine and are analyzed and assessed in some other investigations. Specific errors are present in the signal formation of the measuring head of a CMM that must also be checked using specific test methods.

Deviation from straightness is tested by using a straightness standard, for example, straightedge, laser interferometer, and optical straightness standards. Their metrological characteristics are tested, calibrated, and certified. A standard straight line is orientated along the movable axis and the probe is fitted to the movable component of the machine in the straight line direction. Deviation from the straight line is determined in the two planes perpendicular to each other. The measuring range for straightness is tested in outgoing and returning strokes. If the measurement is made in multipoint touch mode, there must be at least 10 measurement points in the measuring range. Results of measurement are plotted as shown in

Table 2.3. Measurement of the positioning accuracy of the machines and CMMs

Sequence No (Table 2.1)	Accuracy parameter	Method of testing, means, and diagrams
1. 1.1. 1.2. 1.3.	Coordinate displacement deviation during periodic stops (positioning) in axis xx, and z and polar coordinate $\Delta\varphi$.	Measurement is performed during the periodic stops of the movable component of the machine and comparing the readings of its information-measuring system with those of the reference length measuring system (Figure 2. 7). The position measuring accuracy of the moving component 1 of the machine is tested. As the reference length measure the laser interferometer 2 with the reflector 3 and display unit 8 are used. Another possible version of the reference measure is the linear scale 4 with the microscope 5, and yet another measure can be the linear displacement transducer 7 with the movable (reading) head 6. The components of these systems are fitted to the moving component 1 of the machine. The information from the measuring system of the machine is input to the interface unit 9, the output from the reference measure also is input to this unit, further it is controlled by a PC, unit 10, and the results are presented in the registration unit 11 in the desirable form – a graph, diagram, or protocol of the results. Testing of the rotary table is performed in the same way, only the reference measure here is a polygon 2 with the autocollimeter 1, and the rotary table with the drive is indicated as 3 (Figure 2.8).

Evaluation of the results of measurement

The result of measurement is expressed as the difference between the readings of the information—measuring system of the machine and the readings from the indication unit of the reference measure, assessing a systematic error component for reference measure and uncertainty:

$x_k = x_i - (x_{ei} + x_{pi})$; here x_i—i^{th} is the result of an indication unit of the machine; x_{ei}—i^{th} is the result of the reference measure; x_{pi}—i^{th} is the systematic error correction.

Table 2.4. Measurement of straightness of each axial movement

Sequence No (Table 2.1)	Accuracy parameter	Method of testing, means, and diagrams
2.	Deviation from straightness of motion of:	Measurement of the deviation from straightness of the moving component 3 (Figure 2.11 "a," "b") of the machine is made by using the reference measure 2 (straightedge "b," strained wire device "a") and the measuring instrument 1(microscope "a," reflecting mirror fit to the moving component, gauge "b") interacting with the reference measure, which was previously aligned according to the direction of movement of component 3.
2.1	x, in plane xy; $\Delta x(xy)$;	A diagram of measurement from straightness using the laser interferometer was shown in Figure 2.9.
2.2	x, in plane xz, $\Delta x(xz)$;	The measurement in the vertical plane yz is shown in Figure 2.12 "a," and measurement in plane yx
2.3	y, in plane xy; $\Delta y(xy)$;	is shown in Figure 2.13 "b." In both the designations are: 1—laser, 2—remote interferometer, 3—
2.4	y, in plane yz; $\Delta y(yz)$;	double reflector, 4—moving component of the machine, 5—control unit of the laser interferometer,
2.5	z, in plane xz, $\Delta z(xz)$;	6—PC, 7—registration unit, 8—indication unit of the linear transducer of the moving component,
2.6	z, in plane yz, $\Delta z(yz)$	9—control unit of the drive.
		Two different configurations "a" and "b" of the laser interferometer system are used for measurement in the vertical and horizontal planes. The control unit of the laser interferometer evaluates a difference between the two paths of the laser's light beam. The read-out represents the value of an angular movement of prism 3, so it is transferred into the deviation from the straightness of the moving component 4 of the machine. The result is presented in the desired format in the register device.
		The same system of interferometer supplemented by the optical prism as the standard of squareness is used for measurement of perpendicularity of movement.

Evaluation of the results of measurement

The result of measurement is evaluated by enclosing the measured values between two parallel lines (Figure 2.9) and determining the shortest difference between them. The systematic component of the reference measure can be eliminated by using an inversion method as described in the section below.

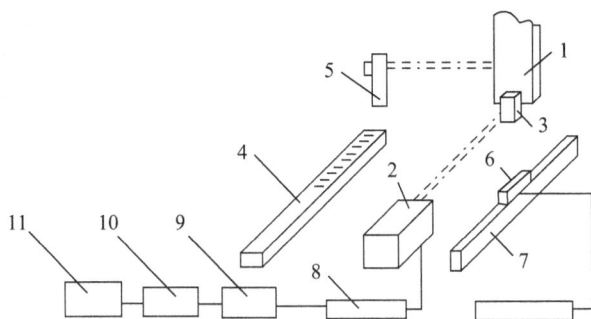

Figure 2.7. Linear positioning accuracy testing.

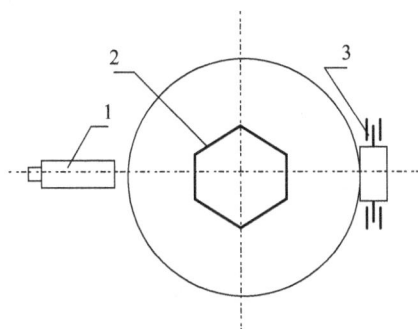

Figure 2.8. Angular positioning accuracy testing.

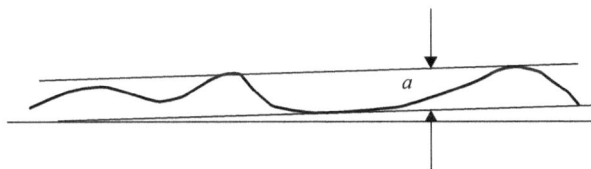

Figure 2.9. Straightness measurement graph.

Figure 2.9. The minimal value a between two parallel lines shows the value of deviation from straightness in micrometers. This value should not exceed the value given by the machine's producer.

2.2.2 INVERSION METHOD IN MEASUREMENTS

An inversion method can be used for better accuracy of measurements. This method includes two measurements on the same reference measure, only accomplishing this at opposite sides of the measure (Figure 2.10). The straightness measurement uses straightedge 2, probe 1, and the movable component 3, of which the straightness of displacement is tested. The reference surface of the straightedge is marked by a dashed line. After performing the measurement according to diagram "a," the position of the straightedge is changed, reversing it according to its horizontal line turning by 180°. Then another set of measurements is performed. (1′) and (2′)—the same elements as in diagram "a" after a reverse of the reference measure. The graphs of the measurements I and II are plotted along the traverse axis x. The true value of the measurement is determined by summing the values of those two graphs and dividing them by 2. The deviation from straightness will be expressed as

$$\delta_{x(yx)} = \frac{1}{2}(\delta_{x(yx)i} - \delta'_{x(yx)i})$$; here $\delta_{x(yx)i}; \delta'_{x(yx)i}$ are the readings of the probe during the measurement according to the plotted curves (I and II).

An inversion method can be applied to other measurements, such as perpendicularity, parallelism, and coordinate positioning. The inversion method is partly used for measurement of rotary table axis run-out, when the reference measure, a spherical substrate, is turned into an angular shift around the same vertical axis. Then the roundness measurement is performed again and the error is assessed on the plotted graph taking account of the phase shift of the reference measure. An inversion method is also used in coordinate positioning measurements, when after the straight measurements the reference linear scale is turned by 180⁰ and measurements are repeated as from the other end of the scale. By summing the

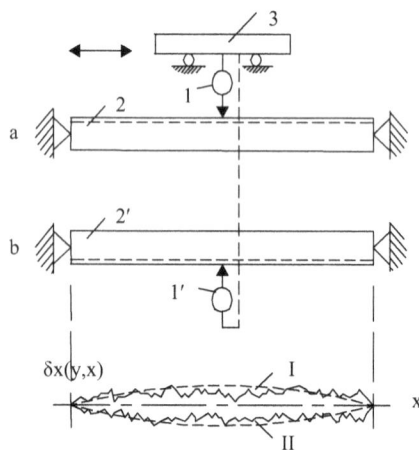

Figure 2.10. Inversion diagram of the reference measure.

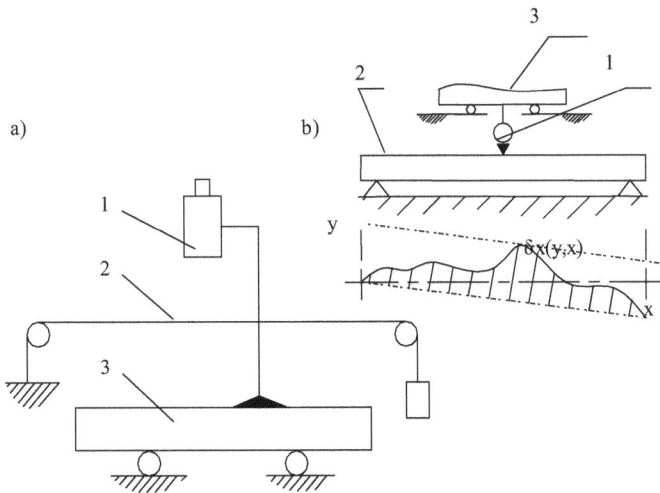

Figure 2.11. Testing the straightness accuracy.

values of both measurements there is a possibility of eliminating the systematic errors of the reference scale.

Two diagrams for straightness testing are shown in Figures 2.11 and 2.12. "a" and "b". Figure 2.11 "a" shows an arrangement for testing the straightness of movement of the part of machine 3 along its coordinate axis, the reference measure is wire 2 under tension being fixed outside of the moving part. The test line of the moving part and the line of the wire are aligned in one parallel direction. The microscope 1 is fixed to the moving part and is directed to the line and level of the reference wire. A deviation from the straight line movement of the part is determined as the difference between the value measured by the microscope and the line of the wire in the perpendicular plane to the axis of the wire. Figure 2.11 "b" shows a similar arrangement using a straightedge 2 as the reference measure and a reflecting mirror 1, fitted to the moving component.

The same parameter is measured by means of a laser interferometer as shown in Figure 2.12. Drawing "a" shows the measurement in the vertical plane (yz) and "b" shows the measurement in the horizontal plane (xy). The main parts of the arrangement are: 1—laser head, 2—optical laser beam splitting cube, 3—reflecting mirrors (pentaprisms), 4—moving part of the machine, 5—laser display and control unit, 6—plotter, 7—printer, 8—display unit of the LDT, 9—drive with its control unit.

The main feature of angle deviations measurement by laser interferometer is registration of the difference between two laser beams produced

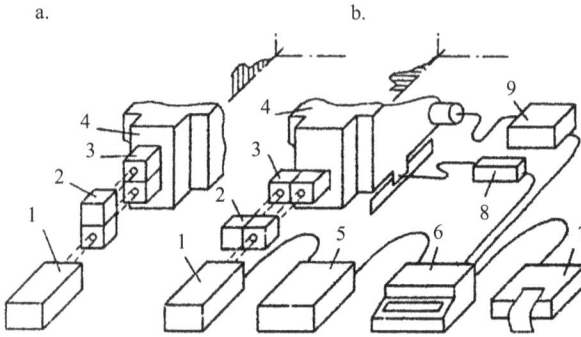

Figure 2.12. Measurement of the deviation from straightness using the laser interferometer.

from one laser source by the optical split system. The light paths from two light beams returning from the reflectors differ due to the angle deviations of the machine's part moving along the coordinate axis. Such an arrangement can be used for the measurement of pitch and yaw parameters. Different means are needed for roll measurement.

Measurement of deviation from perpendicularity of motion in each axis is much in common with straightness measurements. It is measured by using a perpendicularity standard, such as a square, laser interferometer, optical standards (prism, polygon) of perpendicularity. One straight edge of the standard is orientated along the one axis and the probe is fitted to the movable component of the machine. Deviation from the perpendicularityis determined by measurement of deviation from the straightness on the other edge of the standard. If the measurement is performed in multipoint touch mode, there must be at least 10 points measured in the measuring range. The results of measurement are plotted and evaluated as shown in Figure 2.13. The angular value b per reference length (usually, per 200 mm or 400 mm) between two parallel lines inside which the second graph of measurement lies and the perpendicular line to the first aligned direction shows the deviation from perpendicularity in seconds of arc. This value should not exceed the value given by the machine's manufacturer. Deviation from perpendicularity of motion of x axis and motion of y axis; of motion of y axis and motion of z axis; of motion of x axis and motion of z axis are determined. Measurements in a horizontal plane are performed at a height 100 mm from the surface of the machine's table.

The measurement of angular deviations is shown in Figure 2.14. A rotary encoder 2 is fixed to the moving part of machine 1. The axis (shaft)

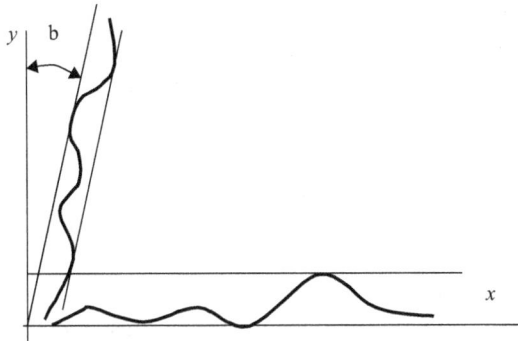

Figure 2.13. The graph plotted for the perpendicularity evaluation.

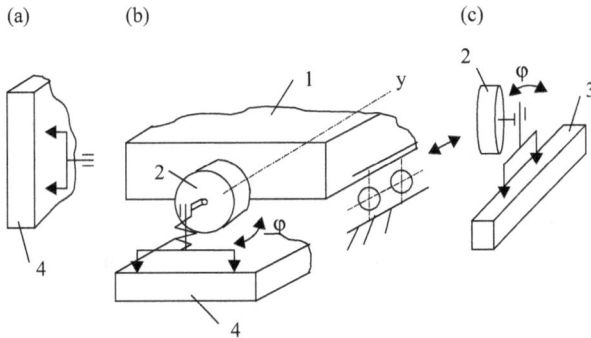

Figure 2.14. Measurement of angle deviations: "a"—pitch deviation in plane xz moving along axis y; "b"—pitch deviation in plane xz moving along axis z; "c"—pitch deviation in plane yz moving along axis y. 1—moving part of the machine, 2—rotary encoder, 3—standard of straightness, 4—standard plate, φ—angle deviation to be measured, 5—double tip sliding perpendicularly to the axis of the rotary encoder.

of the encoder is connected to the double tip which slides perpendicular to the axis of the rotary encoder. Every parallel movement of the double tip does not influence the readings of the rotary encoder. The readings change only in the case when the double tip changes its angular position. In this case the axis of the rotary encoder is slightly rotated on the angular value where the moving part of the machine performs a roll movement according to the standard plate 4 of very high precision of flatness.

The same principle of measurement is accomplished by using two pickups (Figure 2.15) fixed to the moving part instead of the rotary encoder.

The readings from the pick-ups are supplied to the control unit 4, where the readings are transformed into a differential reading showing only the difference of readings between the two pick-ups. This difference is the result of angular displacement (roll) of the moving part during the movement along the chosen coordinate axis. When keeping the distance between pick-ups equal to 412 mm, the readings in μm coincide to values in sec of arc ("). Such a measuring system is easy to arrange in every coordinate axis to control the angular displacement of the moving part.

Angle positioning accuracy of rotary tables or other rotational parts of the machines is tested using the reference measure of angle, polygon with autocollimator, rotary encoder or circular scale of high precision (Figure 2.8). The measurements are taken in at least 12 angular positions in the circle. The rotary table shall be rotating in clockwise and anticlockwise directions. The systematic error of positioning of the rotary table is evaluated as the maximum value of deviation between the readings from the indication unit of the rotary table and the readings from the material substrate or from the instrument (autocollimator, microscope) interacting with the material substrate. Using the reference semisphere of high accuracy (run-out about 0.05 μm), the axial and radial run-out errors are measured. Uncertainty values are determined using the methods of mathematical statistics. Instrumental, graphical, or software (if available) means are applied for eliminating the systematic error before the uncertainty determination.

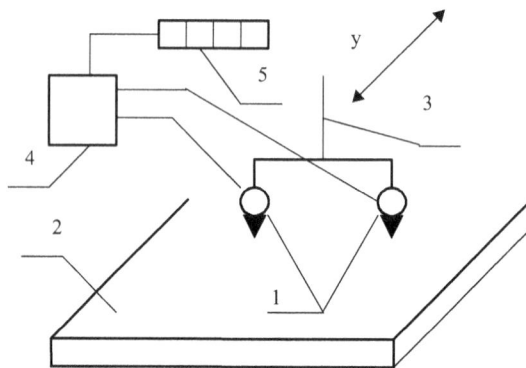

Figure 2.15. Roll measurement of the machine's part (3) moving along axis Y: 1—pick-up, 2—etalon plate, 3—machine's part under control, 4—control unit of the pick-ups, 5—display unit.

2.3   TESTING OF ACCURACY PARAMETERS OF THE ROTARY TABLE

Rotary tables are often used for metal cutting tools and CMMs to extend the manufacturing or measurement capabilities of the equipment. Accuracy parameters of rotary tables are tested in the same manner as linear positioning parameters are performed. Some differences are in the joint assessment of the parameters of the rotary table in connection with the working volume of the CMM. The component to be measured is placed on the surface of the rotary table, so the height and radius of the measurement location lead to a displacement of the measured position on the workpiece. The relationship between the position of the workpiece on the table and the measuring volume shall be tested in the axial, radial, and tangential directions. These relationships are described as presented in VDI/VDE 2617 Blatt 4/Part 4 [17].

Tangential deviation of locus is expressed as

$$O_1 \leq F_1 + r\frac{P_w}{206} + \frac{F_1}{206} \; ;$$

radial deviation of locus is

$$O_r \leq F_r + h\frac{P_t}{206} \, ,$$

and axial deviation of locus is

$$O_a \leq F_a + r\frac{P_t}{206} \, ,$$

where

h: measuring height above the rotary table reference surface in mm;
r: measuring radius referred to the axis of rotation in mm;
F_r: limit of deviations due to axial movement in µm;
F_t: limit of wobble in seconds;
F_w: limit of the angular position deviations in sec of arc;
2.06: conversion constant for the angles.

An influence of errors of the rotary table is assessed according to special methods. For this purpose, two spherical bodies of high accuracy are fitted on the surface of the rotary table at opposite positions according the center of the table. One of them is on the background of the table, the other one is put on some height from the surface. The table is rotating for several times (13 times into the various positions) and the position of the centers of the balls is measured. The error of the rotary table in the working volume of the CMM is calculated as the difference of the real and theoretical positions of the balls' centers.

CHAPTER 3

MEASUREMENT AND MONITORING OF THE ACCURACY OF SCALES AND ENCODERS

The use of linear and circular raster scales for length and angle measurements are first discussed. Traditional methods for angular scales calibration using various angle reference standards are reviewed and some new methods for circular scales calibration are then introduced, including the use of Moiré and Vernier fringe patterns.

The measurement of linear scales and transducers are then discussed and comparators for the calibration of linear standards are described. This is then extended to comparators for angular standards, including the application of piezoelectric actuators for angular rotation. A multipurpose test bench for angle calibration is introduced.

The final section of the chapter discusses techniques for the calibration of geodetic instruments used for angle measurements and a new approach to vertical angle calibration is proposed.

3.1   STANDARDS OF MEASUREMENT OF LENGTH AND ANGLE

Circular and linear scales and raster scales are used in a wide variety of information measuring systems (IMSs) of automatic machines and in length and angle measurement devices. The accuracy of these machines is very high, better than one second of arc per revolution for angular measurements and one micron per meter for length measurements. Such accuracy parameters are used as an indicator of the accuracy level and for the technical specifications for new machines and for in service machines. It is difficult to ensure the accuracy parameters for machines after some years of service and this is why accuracy investigations and their improvement are very important for the creation of new equipment and for the industry as a whole.

The scales are used in control systems with an optical readout, or they are incorporated into linear or rotary encoders for the determination of the position of a cutting tool or the tip of a touch probe for the measurement of machined parts, for example. In this case they are used as a constituent part of the IMS of the machine. The accuracy of the position of the strokes or lines in the raster scale essentially determines the accuracy of the scale and, subsequently, the accuracy of the machine into which it is incorporated. There have been many investigations into the operation of the accuracy parameters of machine tools or measuring equipment but less has been published in the field of optical scale position accuracy. At the same time significant developments in optical sensors, as a means to use fiber optics to assure high accuracy and remote operation in measurement, have taken place. An analysis of the methods and means of calibrating the scales shows that there are still some problems to be solved in this area. One of the most important tasks is the enhancement of the flow of information gained from the measurement process. A second concerns the importance of the sampling process for the performance of the calibration. The measurement of the scales requires a number of points with the pitch of the measurement selected along the full circumference of the circular scale or along the length of the linear scale. These parameters are closely connected to each other. It is impossible to calibrate each stroke of a raster scale using conventional means of calibration, as there are no standards of measurement for each pitch of the raster scale when the pitch of the strokes, for example, is equal to 20 μm for the linear scale, or from 1′ to 5′ (min. of arc) for the circular scale.

Linear and circular raster scales are produced using linear and circular dividing machines using modern techniques involving photo lithography and vacuum technology. For a long time mechanical machines were used for this purpose, where mechanical angle standards were used for circular dividing or angular position determination purposes. In most cases, a worm-glass gear was used for circular dividing machines. The accuracy of the angular position fixed by means of these devices is about 0.3″–0.5″. The "run-out" error of rotation of the axis in such machines is within 0.1 μm. A master lead-screw was used in linear dividing machines for the production of the linear scales. Raster scales, also produced by this equipment, serve as the main element for rotary and linear transducers of high accuracy. The transducers are used as a reference measure for the IMS of metal cutting tools, robotics, and measuring equipment, such as coordinate measuring machines, instruments for navigation purposes and geodesy, and so on.

The angular standard measure, the radian, has not been realized as a standard unit until recently. Nevertheless, the unit of angular measurement in degrees is realized by using the geometric measure of length, in geodesy, in machine engineering and other branches of industry, by multiangular prisms, polygons with an autocollimator, rotary tables, circular scales, and so on. It must be noted that these angle measures are calibrated against higher level measures (etalons) only at several intervals depending on the number of sides (angles) of the polygon or other standard measure taken as an etalon. At the same time, geodetic instruments, rotary tables of metal cutting tools and instruments, rotary encoders, and so on, have a great number of discrete values and the values between the calibrated points remain unknown during the calibration process. Calibration equipment capable of selecting significantly more information would be essential for better accuracy assurance of instruments used in machine engineering, geodesy, building structures, and so on. The theoretical and technical background for the justification and development of such equipment for angular accuracy calibration is presented later in this chapter.

Circular scales were produced for a long time using mainly circular dividing machines. The production of the circular scales is straightforward and is linked with the methods and means of angle measurements. Modern technologies are used for multiplying the original scales made on the dividing machines by means of photo lithography using well-known technologies, such as Aurodur, Diadur, and so on. These methods have not altered much in time and are especially valid for circular scales measurement. High accuracy and discretion of circular scales measurement remain an actual problem even nowadays.

Technical measuring equipment, instruments, and systems, such as linear and circular scales, measuring transducers, and numerically controlled machine tools require an IMS that provides measurement information or information about the position of part of a machine or instrument. A typical measuring system consists of linear and rotary transducers (encoders) whose measuring part is fixed to the moving part of the machine and the index part, the measuring head, is fixed to the base of the machine. Using such systems, it is difficult to measure long strokes of the machine, as the accuracy for a long measuring translational transducer is much more difficult to achieve in comparison with a short one. Accuracy characteristics can be analyzed in terms of the informational features of a short measuring transducer (for example, 400 mm) comparing it with a long one (for example, 2000 mm). Usually, a modern machine has a measuring system with a resolution equal to 1 μm, so in these two examples the indicating unit of the machine will show 4×10^5 or 2×10^6 digits. Examples of such systems could be measuring laser interferometer systems or the display units of CMMs. The accuracy parameters associated with some popular angle standards are summarized in Table 3.1.

A modern method of angle measurement has been developed using the "ring laser" as a reference measure of angle (Figure 3.1). This method is slowly taking its place in machine engineering due to its dynamic nature, but until now it was most widely used in navigation systems, and so on. Fiber optic gyroscopes (FOGs) are used mainly in navigation instruments, their main feature being the ability to operate in one plane, that is the plane in which all the arrangement is positioned. The device includes a laser source

Table 3.1. Accuracy parameters of some angle standards

No	Angle standard of measure	Resolution of measurement	Standard deviation of the readings	Bias
1.	Polygon—autocollimator	10°; 15°; 24°; 30°,...	0.15″	0.30″
2.	Moore's 1440 precision index	15′	0.04″	±0.1″
3.	Circular scale—microscope	3°, 4°, 5°	0.2	~3″
4.	Photoelectric rotary encoders	1″; 0.1″	~0.3″	~1″
5.	Laser gyro	0.1″; 0.01″	0.05″	<1″

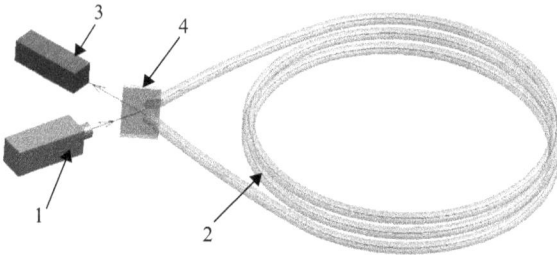

Figure 3.1. The fiber optic gyroscope "ring laser": (1) light source (laser), (2) optic light guide, (3) light receiver (photodiode), and (4) optic prism.

and an optical prism, and the resulting split laser beam propagates in oppo-site directions. The angle measurement is performed by the comparison of the split beams phase difference, in such a way that a very high precision is being achieved. When there is a need to measure an angle displacement in 2D directions, then two FOGs perpendicular to each other are used and consequently, for 3D measurements, three FOGs must be applied. The FOG measures speed up to about 5000 rev/min with an accuracy of ≈1%.

3.2 MEASUREMENT OF CIRCULAR SCALES

Measurement of circular scales has its own specific problems. Additional features that help to perform this measurement is the fact that the sum of the full circumference is always equal to 360°. This enables the cali-bration of circular scales using one reference angle throughout the whole circumference and thus the errors of angular values may be analyzed as the difference between the real values of reading devices and the values of the calibrated reference angle. In the metrology of circular scales that was mostly developed in geodesy and astronomy, there is terminology used such as "error of the scales diameter".

The expression of *diameter* means the line going through the strokes lying at the opposite sides according to the center of the scale. In most geodetic measurements the errors of diameters or the errors between the diameters are determined. It helps to avoid errors due to the eccentricity of the scale to be measured and due to the trajectory of axis rotation. The error of diameters can be expressed in terms of the algebraic sum of errors of two opposite strokes:

$$\Delta(\varphi_\Sigma^i) = \frac{1}{2}\left(\Delta\varphi_\Sigma^i + \Delta\varphi_\Sigma^{i+180^\circ}\right),$$

where $\Delta\varphi_\Sigma^i$ and $\Delta\varphi_\Sigma^{i+180^\circ}$ are the errors of angle position of two opposite strokes.

The errors of circular scales are determined by various methods approved in the written standards:

- the method of approximation;
- the method of opposite matrix;
- the method of Yeliseyev (or Heuvelink);
- the method of Wild.

These methods are supported by the written standards, although different organizations are used to apply different methods of measurement.

The following methods of angular scales calibration are used in machine engineering and instrumentation:

- the comparison of the angular values of the scale strokes with the values of the reference scale or other reference measure of angle;
- the comparison of the angular position of strokes of the scale with the reference angle created by the strokes of the same scale. This method is also called calibration with the constant angle in the full circumference.

The methods of circular scales calibration were created and developed by such famous scientists as H. Bruns, G. Schreiber, A. Perard, H. Wild, H. Heuvelink, S. Yeliseyev, and so on. The diameters errors are determined by using these methods at angle intervals, applying equal control angles, for example, 10°, 20°, and 30°. Processing the diameters errors by the Bruns method, linear equations of the strokes of scale position are created. The number of equations is equal to the number of angular position errors to be determined. For example, by measuring at every 3°, a system of 60 equations is created. The Yeliseyev (or Heuvelink) method was a further development of the former method, simplifying the number of calculation operations. The diameters errors are calculated determining their mean arithmetic values, and for the enhancement of accuracy of the calculations error weight parameters are introduced. Thus, the error evaluation is stochastic; it differs from the real values of the relevant errors of the diameters. If the pitch of measurement of the circular scale is not small enough, so the discretion of the stroke errors is not big enough; that is, the error is determined at quite large intervals of the scale. This is the reason why comparative scale measurements are performed in machine and instruments engineering by using other angle standards with a much higher discretion of reference angle measure.

The diameter errors by comparison methods were determined from the period of the early development of geodetic and astronomic instruments. Thus, n equations for measurements performed at a chosen pitch of the scale measurement, for example, 10°, are designed. It is accepted that $(\varphi) = 0$, at the beginning of measurement, and the other errors are calculated assuming that the errors are distributed evenly in the circumference. By making this assumption, a value $\dfrac{1}{n}\sum_{i=1}^{n}\omega_i$ is subtracted from the results received, where ω_i denotes the errors of corresponding opposite strokes. Such an approximation is used by many authors, including the Yeliseyev method. Nevertheless, it is not accurate as the errors of different strokes and errors of the position of diameters are not equal and their sum is not equal to zero.

According to the main methods of scales calibration, for the calculation of scale errors, it is necessary to associate the positions of strokes on the scale opposite to the constant angles. For example, to determine the error at every 5°, it is enough to calibrate the scale applying the constant angles of 40° and 45° (or 20° and 45°). The constant angles for calibration are chosen as repetitive values of the desirable interval of calibration.

Some of the angular calibration processes are performed using Moore's 1440 Precision Rotary Index Table (Figure 3.2) as the angle standard and an angle polygon prism of 12 sides with an autocollimator. Moore's 1440 Precision Index is an angular measuring device consisting of two serrated plates joined together to create the angle standard of measure. During the measurement the upper disk of the index is lifted, the lower part rotates with the object to be measured, after that the upper part is lowered back and the measurement of the angular rotational error is performed by the autocollimator. The repeatability of the readings of the autocollimator do not exceed ±0.02″. The accuracy of the axis of the index rotation is in the range of 0.1 μm, the interval of angle measurements being 30° (in the case of using the 12-sided polygon). Every position is usually repeated about 10 times, the values of polygon calibration being within the limits (−1.5″ and +2.5″).

The rotary test table, the standards of angle, and measuring instruments are shown in Figure 3.3. The standards of angle as shown in Figure 3.3 are the most widely used standards in machine and instrument engineering. The items shown in the picture are: (1) base of the angle-measuring equipment, (2) axis of rotation, (3) worm wheel, (4) circular scale, (5) photoelectric microscope (PM), (6) angle-measuring device

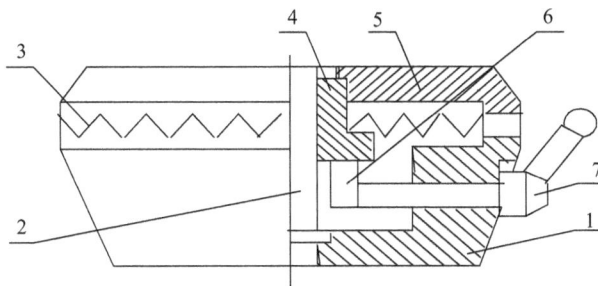

Figure 3.2. Rotary index table: (1) basic part, (2) axis for lifting and rotating, (3) serrated meshing, (4) hub, (5) upper disk, and (6,7) cam lifting device.

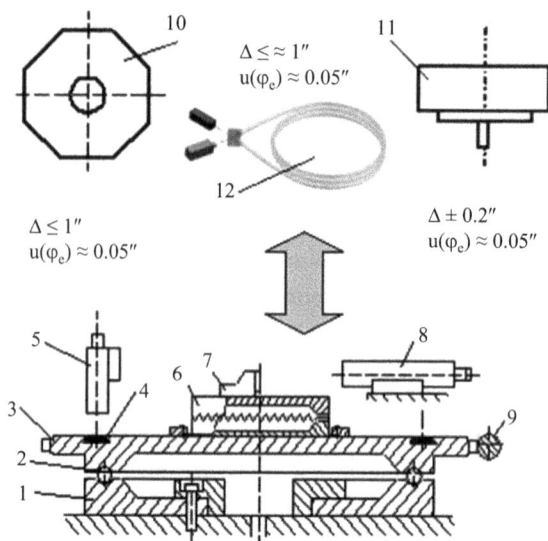

Figure 3.3. Arrangement for angle calibration with different angle standards.

(Moore's 1440 Precision Index), (7) mirror, (8) autocollimator, (9) worm glass gear, (10) multi angle prism (polygon), (11) rotary encoder of high precision, (12) laser gyro (also shown in Figure 3.1). The systematic error value and uncertainty of the standard measure are indicated.

The worm gear drive, step motors, or piezoelectric drives are used to move the object to be measured into the required position. The items 4–5, 6–7–8, 10–11–12, can be chosen as the standard measure for angular displacement control. High-accuracy photoelectric transducers (rotary encoders) are included in the range of measurement standards, together with the most advanced instrument for angular measurements, the laser gyro. When using the circular scale as the standard of measure, two PMs are used to avoid the influence of eccentricity for the angular measurements. The advantages of rotary encoders provide good possibilities to automate the measuring process. Devices using an angle interferometer can be implemented only in laboratory conditions.

The angle reference standards that can be used include the circular scale with the microscope, Moore's Index table and autocollimator, and the other standards as shown in the upper part of Figure 3.3. The multiangle prism (polygon) has a calibration uncertainty equal to ~0.02″ and the high precision rotary encoder has a calibration uncertainty equal to ~0.05″. Mechatronic techniques can also be used and tested in the case of the drives, that is the piezoelectric plate for the table rotation, PMs, and autocollimator with charge coupled device (CCD) cameras installed for alignment and centering of the objects to be measured.

This entire complex can be shown by experiment to be quite effective and progressive for high-precision angle measurements and as the basis for the creation of the flat angle standard.

The most widely used angle-measuring standards are polygons, precision indexing tables, and circular scales. Some other precision measuring devices, such as ring lasers and precision rotary encoders, are also used. Reading instruments, such as microscopes, scanning heads, mirrors, and autocollimators, are used with the angle standards. A high accuracy of the comparator for the flat angle unit transfer was achieved by tests carried out in PTB, Germany. The angular comparator WMT 220 of very high accuracy was used for the calibration of electronic autocollimators for the flat angle unit, radian, and transfer according to the ISO standards. The tests were based on the use of incorporated circular gratings and photoelectric incremental read-out heads of the values measured. This angle comparator is a measuring device for plane angle calibration ensuring an uncertainty of measurement of 0.005″. The comparator is mounted on a solid, vibration-isolated granite plate and comprises an air-bearing rotor of high accuracy with an axial plane bearing and a cylindrical radial bearing. Rotational speed is between 7.5 rev/min and 7.5″/min for dynamic measurements. The angle-measuring system consists of a circular glass index disc with a circular scale of 2^{17} graduation periods (131 072) on a circle, ~400 mm in diameter. Graduations are scanned by eight scanning heads uniformly distributed over the circumference of the scale. After an interpolation of the signal period 2^{30} (1,073,741,824) measurement steps are achieved, which corresponds to an angle-measuring discrete step of ~0.0012″. The rotation of the upper disc was performed on the radial and axial air bearing of the highest accuracy. An uncertainty of 0.007″ was determined as a result of calibration of electronic high-resolution autocollimators in repeating consecutive steps of 0.005″ in the transfer of the standard unit of plane angle. It was stated that calibration in very small intervals that are near to the resolution capability of the autocollimator gives information about the possible short period bias and as a consequence the autocollimator resolution can affect the measurement. The comparator can be self-calibrated against a test system or against other angle-measuring systems applying the so-called rosette measuring method. For the calibration of high-quality precision polygons or incremental angle-measuring systems $U = 0.05″$ ($k = 2$) is achieved.

The raster circular scale can include from several hundred to several tens of thousands of strokes. It is evident that there is no easy means for creating the standard angle for the measurement of each stroke and thus errors of the "short period" are not determined. Some methods developed for solving problems of such kind use a phase shift in measurement. The problem is becoming more complicated by the wide variety of scales used in modern automated engineering. Raster scales of a wide range of accuracy and dimensions, systems of graduation, and so on, are used for special tasks in instrumentation and automation. (360°, 2^n, 10^n number of strokes in the circle, various coded scales, etc.)

3.3 APPLYING NEW METHODS FOR MEASURING CIRCULAR SCALES

A newly developed method for circular scales calibration is based on the application of the 180° (2π rad) angle created during the same measuring process as a standard measure for the calibration of the circular scale. Such a standard can be set with an accuracy not less than 0.1″ of standard deviation in the case of a high-accuracy rotational axis and using high-accuracy PMs. The measuring scheme and the principal process of measurement are as shown in Figure 3.4. A simplified method of measurement is presented and the main task is to demonstrate the algorithm and the means for data processing.

Traceability of angle measurements is based mainly on the standard of the plane angle-prism (polygon) calibrated at an appropriate accuracy. Some eminent metrological institutions have established special test benches (comparators) equipped with circular scales or rotary encoders of high accuracy and polygons with autocollimators for angle calibration

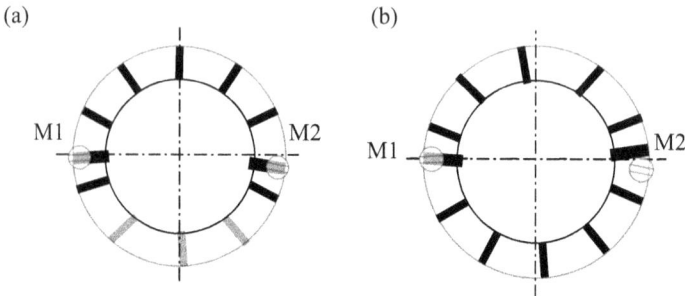

Figure 3.4. 180° angle setting as reference standard for the circular scale calibration: (a) a preliminary microscopes setting on the diametrically opposite strokes of the scale; (b) the same position of the microscopes after the rotation of the scale by 180°.

purposes. Nevertheless, the standard (etalon) of plane angle, the polygon, has many restrictions for the transfer of the angle unit, the radian (rad), and other units of angle. It depends on the number of angles formed by the flat sides of the polygon. A possibility to create the standard with the angle equal to π rad or half the circle (or the full angle) is proposed. It can be created by a circular scale with the rotational axis of very high accuracy and two precision reading instruments, usually, PMs, placed opposite when using the special alignment steps. A great variety of angle units and values can be measured and its traceability ensured by applying a third PM on the scale. Calibration of the circular scale itself, and other scales or rotary encoders as well, is possible using the proposed method with an implementation of π rad as the primary standard angle. The method proposed assures the traceability of angle measurements at every laboratory having the appropriate environment and appropriate reading instruments, together with a rotary table with a high-accuracy rotation axis, that is, the rotation trajectory (run-out) is in the region of 0.05 mm.

The angle of $180°$(π rad) of high precision can be set up by means of the circular raster scale to be measured, two opposite microscopes and the axis for the scale rotation (Figures 3.4 and 3.5). Two microscopes M1 and M2 are set on the diametrically opposite strokes of the scale, Figure 3.5 (a). After the rotation of the scale by $180°$, the strokes of the scale will take a position as shown in Figure 3.5 (b). By adjusting the microscopes M1 and M2 in the tangential direction, a position can be reached when the readings from both microscopes are equal in the absolute value after the rotation of the scale to $0°$ and $180°$. Therefore, microscopes M1 and M2 are set for the determination of the standard angle of the measurement, equal to $180°$ or π rad.

The measurement of the angle errors between the strokes of the scale is performed by setting a third microscope on the stroke at the chosen step

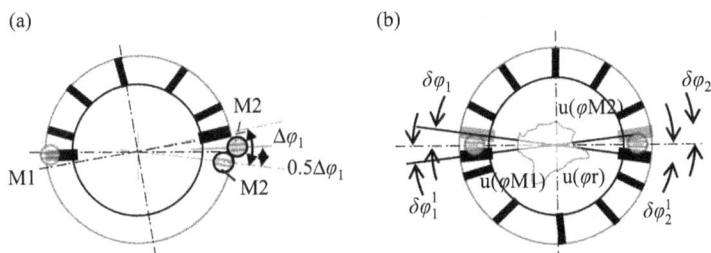

Figure 3.5. Setup of the microscopes: (a) determination of tangential displacement of the microscope M2; (b) position of the microscopes at the primary standard angle of $180°$.

of measurement (Figure 3.6). The third microscope is set on the stroke of the scale at the angle φ_t. The scale is moved by angular steps in the clockwise direction, the measuring data being registered as a_i and b_i, where $i = 0,1,2...,(2n-1)$. a_i and b_i are the data measured by the first and second microscopes; $2n$ is the number of strokes in the scale.

The position as shown in Figure 3.6 is: $|a_i| = |b_i|$. The initial presumption is taken that $a_0 = 0$; $\delta\varphi_0 = 0$; $\delta\varphi_n = b_n$; $\delta\varphi_t = \Delta$; where Δ is the error of the angular position of the stroke with the index t; φ_t is the angle of the third microscope from the "0" point; and $\delta\varphi_i$ is the errors of angle of relevant strokes of the scale. The result of measurement is expressed by a system of linear equations:

Indexes of readings b vary according to the number of strokes from n to $(2n-1)$; indices of a follow the strokes with the numbers $i = 0, 1, ..., t$. The general expression for readings will be

$$\begin{cases} a_i = \delta\varphi_i - \delta\varphi_{i+t} + \Delta; \\ b_{n+i} = \delta\varphi_{n+i} - \delta\varphi_{i+t} + \Delta. \end{cases}$$

After performing the measurement for the full circumference, and by summing both sides of equations, it will yield

$$\sum_{i=0}^{2n-1} a_i = \sum_{i=0}^{2n-1} \delta\varphi_i - \sum_{i=t}^{2n-1+t} \delta\varphi_i + i\Delta;$$

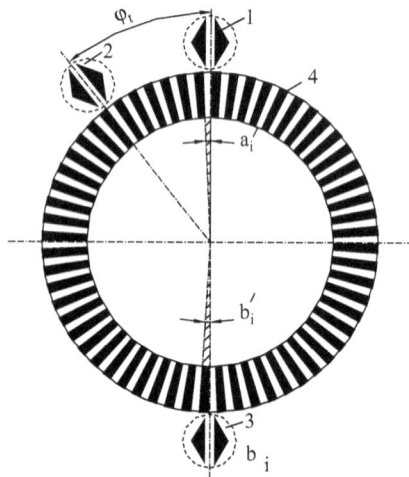

Figure 3.6. The diagram of measurement by determining the 180° standard of angle.

$$\sum_{i=0}^{n-1} b_i = \sum_{i=n}^{n-1} \delta\varphi_i - \sum_{i=t}^{2n-1+t} \delta\varphi_i + i\Delta \; .$$

For the first equation: $\sum_{i=0}^{2n-1} \delta\varphi_i - \sum_{i=t}^{2n-1+t} \delta\varphi_i = 0$, then

$$\Delta = \frac{1}{i}\sum a_i .$$

Furthermore, the equations can be solved in respect of either a_i or b_i. Therefore, it must be noted that various possibilities are available for the determination of the diameters errors using the expressions derived. Here the calculations will be demonstrated according to the formula involving expressions of a_i. Theoretically, it is possible to measure and determine the errors of every stroke in the circular scale, albeit the great number of strokes that are present. This is particularly important nowadays, as the circular scales are made of very small diameter having thousands of strokes in the scale. It is impossible to perform such measurements by using the existing circular scale calibration methods. Applying an axis of rotation based on aerostatic bearings and using microphotoelectric reading devices, a real possibility for its implementation exists. The error determination functions, consisting of a great number of equations, also lead to problems as a specialist mathematical software package will be required. As an example by measuring the scale at every 1°, an equation system of 360 members will be created; by measuring the scale at every 1/3 degree there will be 1,080 members, and so on. A raster scale having 21,600 strokes will require the same number of equations to be solved for the error determination.

Operations with more data do not introduce anything substantially new; only some parameters in the instruction code for a mathematical software package need to be changed. After the solution process, which takes noticeably longer, it makes sense to output only the final results in the row or table format as instructed. The capabilities of calibrating circular scales can be validated by computer modeling described in papers presented in the reference list at the end of the book [18–23].

Experiments were performed to compare the half circle method with the conventional method, that is, calibration with the Index table 1440. The results are shown in Figure 3.7 (where graph 2 is the half circle method and graph 1 is the Index table 1440 method). These results show quite a good agreement, including the extreme points a, b, and c, where the errors are the largest.

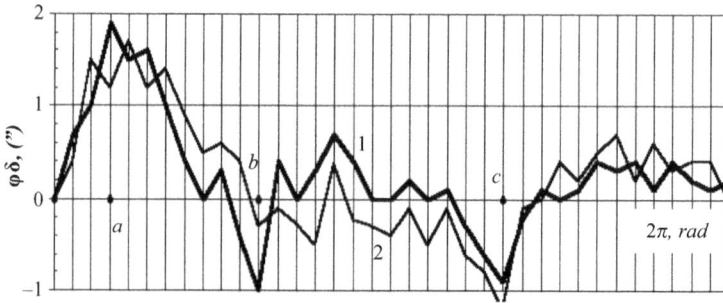

Figure 3.7. Comparison of measurements, performed by using the rotary index table (1) and the method of the half circle (2).

The strokes symmetry method: This method is very effective for simplifying the measurement and for the automation of the process. The diagram explaining the measurement is shown in Figure 3.6. The difference from the measurement described previously is that readings are taken only from the third microscope, at the moments when the positions of the opposite strokes under the microscopes (1) and (2) are of equal value and have opposite signs: $|\delta\varphi_i| = |\delta\varphi_{n+i}|$. The readings from the third microscope are designated as c_i, and the initial reading (when the stroke "0" is under the first microscope) will be

$$c_{2n-1} = \delta\varphi_{2n-1} - \frac{1}{2}(\delta\varphi_{2n} + \delta\varphi_n).$$

Further readings give the system of equations:

$$\begin{cases} c_i = \delta\varphi_i - \frac{1}{2}(\delta\varphi_{i+1} + \delta\varphi_{n+i+1}) + \Delta, \\ \Delta = -\delta\varphi_{2n-1}, \end{cases}$$

where $i = 0,..., 2n - 1$.

The system has $(2n + 1)$ equations and the same number of the unknown members.

The general expression of the equations of the errors will be

$$\begin{cases} \delta\varphi_i = \frac{1}{2}[\delta\varphi_n + \delta\varphi_0 - C_{n+i} + C_i - \sum_{j=n}^{n+i-1}(C_j + C_{n+j}) + 2i\Delta], \\ \delta\varphi_{n+i} = C_{n+i} - C_i + \delta\varphi_i. \end{cases}$$

In such a manner all the errors of the circular scale can be determined. The calculation program was developed in the FORTRAN language and the final solution of the measurement results is given, when the readings from the scale are taken at a small rotating angle, not coincident with the line between the two opposite microscopes and the third microscope.

Examples of change from discrete to analog information in the measuring system can be presented by angular accuracy calibration of the raster scales. The calibration of the scale stroke-by-stroke (at a predetermined pitch) gives discrete calibration information. A possibility to transfer it to the analog form is shown in Figure 3.8. It is performed by creating a Moiré (Figure 3.8, a) or a Vernier (Figure 3.8, b) fringe pattern, one of the scales serving as a reference. By using the expressions for determination of the light transmission function Π, it is possible to calculate this and the modulating function m in the period of the pattern.

These functions are

$$\Pi = \frac{1}{W}\sum_{j=1}^{i-1} a_j = 0,25$$

and coefficient of optical modulation

$$m = \frac{\sum_{k=1}^{i}(i-k) - \sum_{k=1}^{i}(k-1)}{\sum_{k=1}^{i}(i-k) + \sum_{k=1}^{i}(k-1)}.$$

Here i is the coefficient of optical reduction of optical composition of two raster scales; $k = 1, 2, 3, \ldots a_j$, the width of transparent part of the raster scales composition. W is the pattern period of the Moiré or the Vernier scale.

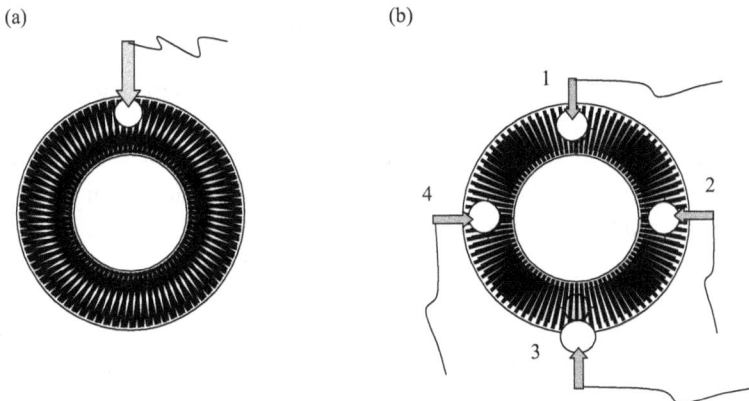

Figure 3.8. Use of Moiré (a) and Vernier (b) pattern for angular accuracy calibration of circular scales.

One, two, or four analyzing gaps with photocells are located on the optical patterns shown in Figure 3.8. For the Moiré pattern it is enough to place one photoelectric sensor (in the upper part of the scales, Figure 3.8, a) on the pattern and rotate it on the table axis of high accuracy. Roundness measuring instruments are used for this purpose. The output voltage from the sensor is proportional to the form of the Moiré fringes, and this is a function of the accuracy of displacement of the strokes of the scale under control.

For the Vernier raster scales two or four sensors are placed on the pattern. Sensors 1, 2, 3, and 4 are placed at angles of 90° to each other. It is evident that after rotation of the scales for every 90°, the output voltage must be equal for raster scales having no errors, and every systematic error of the scale under control will give a bias of the output voltage proportional to the error detected. The same can be achieved by placing the sensors at a distance equal to the Vernier pattern period W or on the spacing, equal to $0.5W$ or $W(k + 0.5)$, where $k = 1, 2, 3$. When the raster composition is rotated under the high-precision rotary table (for example, roundness measuring instrument), the output voltages from the photocells compensate each other. The output voltage from the photocells is amplified, transferred to the summing unit and then passed further to the recording device. It is obvious that in the case of a high-accuracy raster, the output voltage in the recording device shows no change, and the graph will be a straight line (or a sine wave). When a raster pitch error occurs, the fringe pattern moves by a distance equal to the pitch error multiplied by i. The output voltage changes accordingly, providing a clear indication of the systematic error of the raster scale.

The sensitivity of the index sensors used with the photocells can be tuned, enabling a measurement of a wide range of errors to be controlled. The measurement can be performed on various kinds of raster scales using Vernier or Moiré fringe patterns. Precise roundness measuring devices and templates are very relevant for such a purpose and the results can be represented in a digital or graphical form. The method discussed is more informative and efficient than the use of measurement by comparison with the reference angle measure. It can be especially effective for industrial needs and applications. All these applications are to be implemented using every constituent of the mechatronic arrangements mentioned above. Therefore, the "short period" errors are not determined. Bearing in mind, that short period errors in geodesy are considered to be errors within the period of 1°...3°, it can be stated that the quantity of information required for such measurements is very large. The method designed for solving problems of this kind is as follows.

The classical *constant angle setting in full circle method* was modified for circular scale calibration. A single step of measurement applied for the modified method is shown in Figure 3.9.

The systematic errors of each scale stroke could be calculated according to the equations:

$$\delta\varphi_1 = 0,$$

$$\delta\varphi_2 = \left(\Delta M_{2.2} - \Delta M_{1.1}\right) - \frac{\sum\limits_{i=1}^{n}\left(\Delta M_{2.(i+1)} - \Delta M_{1.i}\right)}{n},$$

$$\delta\varphi_3 = \sum\limits_{i=1}^{2}\left(\Delta M_{2.3} - \Delta M_{1.2}\right) - \frac{2 \cdot \sum\limits_{i=1}^{n}\left(\Delta M_{2.(i+1)} - \Delta M_{1.i}\right)}{n},$$

$$\delta\varphi_n = \Delta M_{1.n} - \Delta M_{2.n-1} - \frac{\sum\limits_{i=1}^{n}\left(\Delta M_{2.(i+1)} - \Delta M_{1.i}\right)}{n},$$

where $\Delta M_{1.i}$ are the readings taken by microscope M_1, $\Delta M_{2.(i+1)}$ are the readings taken by microscope M_2, n is the total number of the readings in the full circle.

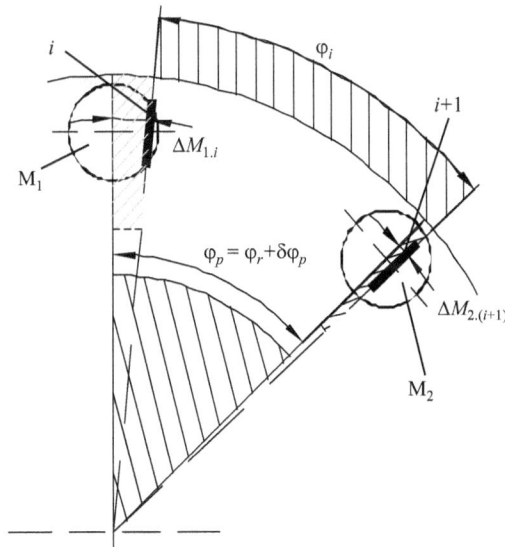

Figure 3.9. Measurement step using two microscopes, M_1, M_2, i is the number of scale stroke ($i = 1, 2, 3, \ldots n$).

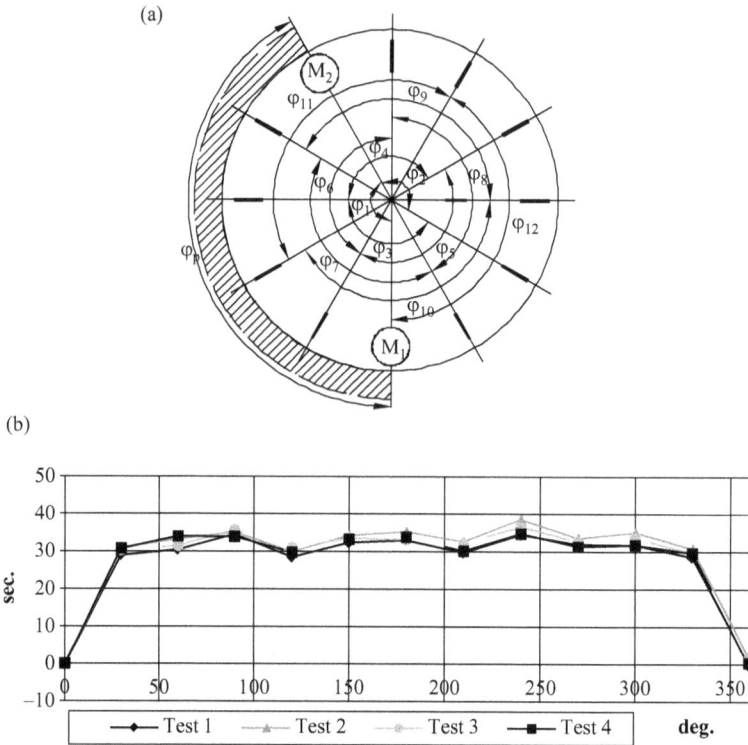

Figure 3.10. (a) Circular scale measurement using an angle not a multiple of 2π (150° in this case). M_1, M_2—the microscopes placed on the strokes of the scale; (b) graphs of the circular scale errors.

The modification allows the calibration of the circular scale without the need for precise mechanical scale positioning. The calibration can be performed at an approximate angular position only placing the investigated stroke in the field of view of the microscope.

In Figure 3.10 (b), the comparison of the rotary table angular position errors determined by multiangular prism/autocollimator and a circular scale/microscope (with systematic errors of strokes compensated) is shown. As can be seen the results are very similar. According to further calculations it was determined that the root mean square error of the angular position determination using a calibrated circular scale and microscope (relating to the multiangular prism/autocollimator measurements as the reference) was not larger than $\sigma = 0.172''$.

After a full cycle of the measurements performed, the real angle between the microscopes ($\varphi_r + \delta\varphi_p$) for each test can be calculated. The approximate angle between the microscopes (φ_r) being 150°, the deviations of angle will be determined from the equation:

$$\delta\varphi_p = \frac{\sum\limits_{i=1}^{n}\left(\Delta M_{2(i+1)} - \Delta M_{1i}\right)}{n}.$$

Each scale stroke bias at each test ($\delta\varphi_i$) with reference to the first (0°) stroke can be calculated using the equation:

$$\delta\varphi_i = \delta\varphi_{i-1} + \left(\Delta M_{2i} - \Delta M_{1(i-1)}\right) - \delta\varphi_p.$$

Judging from the data received (the calculated scale strokes biases for each measurement series) the standard deviation of each bias measurement is quite small (not exceeding $S_i = 0.423"$, including the standard deviation of the modernized microscope measurements $S_m = 0.0125"$). Since the biases were calculated with reference to the first stroke (considering the 0° stroke as being the reference one), the standard deviation of each specific stroke should be determined as

$$S_k = \sqrt{\sum\limits_{i=1}^{n} S_i^2},$$

where k is the number of the scale stroke.

Thus the maximum standard deviation will be reached at the last calculated stroke:

$$S_{max} = \sqrt{\sum\limits_{i=1}^{n} S_i^2},$$

where n is the total number of the calculated scale strokes.

In the case of measurements where the approximate angle between the microscopes was 95°, the maximum standard deviation will be reached at a scale stroke marked as 90° and $S_{max} = 2,621"$. It should be noted that despite quite a high standard deviation of the last stroke bias using a large number of measurements should practically eliminate any kind of random error in the scale bias determination, and since its value needs to be determined only periodically (during the rotary table scale calibration, which should be performed quite seldom) it is quite possible to achieve appropriate calibration results.

The graphs of measurements of angular position errors of the rotary table determined by polygon/autocollimator and circular scale/microscope are shown in Figure 3.11.

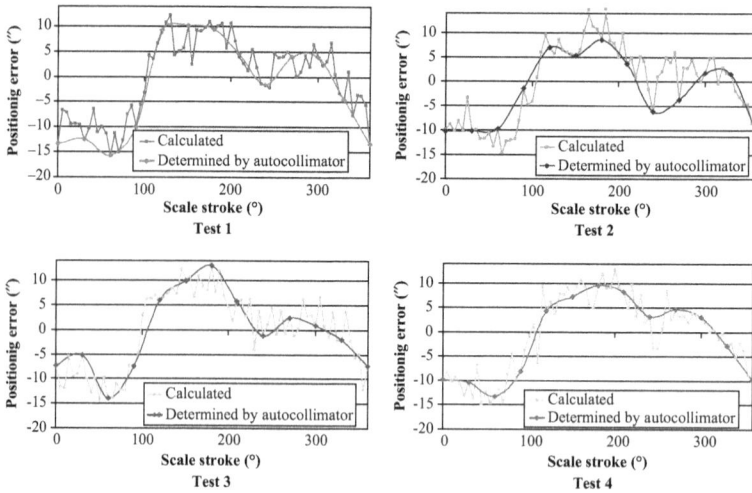

Figure 3.11. Positioning error of rotary table determined by multiangular prism/autocollimator and that calculated by circular scale and microscope.

The graphs show that the angular positions of the rotary table determined by the circular scale and microscope match the position determined by polygon and autocollimator quite well. The standard deviation of the calibration data from the prism/autocollimator measurements is calculated as

$$S_{a/p} = \sqrt{S_a^2 + S_p^2},$$

where S_a is the standard deviation of autocollimator measurements ($S_a = 0.0405''$) and S_p is the standard deviation of polygon calibration ($S_p = 0.1''$). The general standard deviation of the autocollimator/polygon measurements does not exceed $S_{a/p} = 0.108''$. The general root mean square error and the uncertainty of angle position determination at a confidence level of 0.95 using the measuring data are calculated as $\sigma = 0.172''$ and $\varepsilon = 0.0295''$, respectively.

The readings from the microscope scale are calculated in the same way, thus enabling it to be used as a reference measure for further geodetic instrument calibration. Some statistical characteristics are calculated during the evaluation of the results, including an estimate of the standard deviation S at Student's t-test and degrees of freedom n. This is used for the final determination of systematic part (bias) of the angular error with the uncertainty ε. Some results of measurement using an electronic theodolite, Wild the odomat T1000, are presented in Figure 3.12.

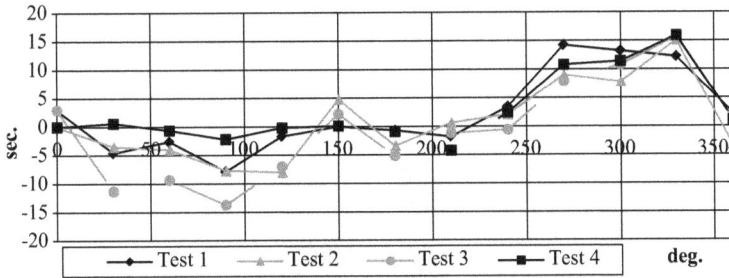

Figure 3.12. Graph of the tacheometer's systematic errors.

The modifications of the method allow the calibration of the circular scale at any angular pitch in completely automated mode (with an approximate positioning). The determination of the systematic errors of every stroke of the scale during the calibration has been performed at an angular pitch of 5° (microscopes positioned at 95° to each other). The readings from the multiangular prism autocollimator were taken simultaneously at a pitch of 30°. The results of the scale calibration are shown in Figure 3.10, (b). As it can be seen there are some significant systematic errors of the scale strokes but since they are stable and already determined, it is relatively easy to compensate them and include the results of calibration in the process of measurement. The rotary table of the test rig can be positioned very roughly using an angle encoder and knowing the number of each scale stroke and having the calibration results (i.e., systematic error) for that particular stroke stored in the PC, the precise angular position of the rotary table can be immediately calculated by the control computer.

Therefore, it may be stated that angular position measurement using the circular scale and microscope can be successfully implemented with a precision close to the one obtained by the multiangular prism and autocollimator but at much smaller pitch. The small pitch of measurement is especially important for testing or calibration of angle-measuring geodetic instruments as well as for any other type of modern instrument for angle measurements (i.e., angle encoders, raster, or coded circular scales, etc.).

The importance of the measurement of circular scales of small diameter is illustrated in Figure 3.13. The diagram shows that the errors measured have a smoother character of distribution, when their diameter is larger and the diagram has sharp peaks and curves of the scales, when the diameter is small.

The technical specification parameters mentioned above would suggest that there is a technical possibility to create an angle calibration bench

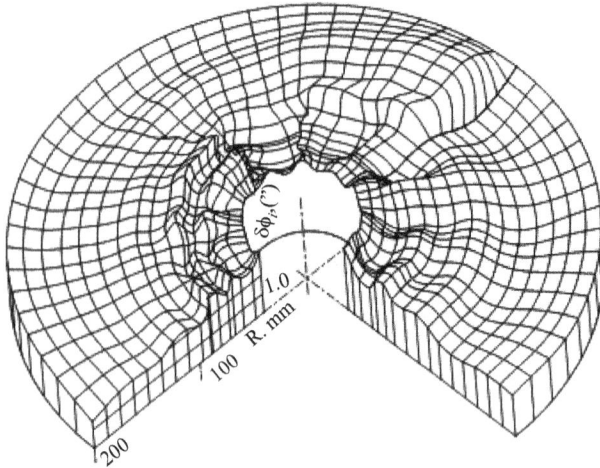

Figure 3.13. Diagram of multiple measurements of circular scales with various diameters.

of very high accuracy. Such equipment could also be used for the calibration of geodetic instruments and machine engineering devices. It is possible to calibrate the angle-measuring devices to an accuracy equal to ~3″. The discretion of the calibration is also easily achieved, and such features provide the possibility to calibrate angle standards and instruments which have 10^n or 2^n strokes or signals per revolution.

A novel idea for the measurement of linear and circular raster scales is by the determination of a bias from a straight line and from a circle, respectively. This idea is implemented by applying a Moiré fringe combination technique and the modification of the resulting Moiré fringe parameters, expressed by discrete functions.

Linear and rotary transducers and encoders using raster scales have two-raster scales in combination to achieve the light transfer modulation. Many transducers and several means of measurement have been devised on the basis of the Moiré fringe pattern formed by two raster scales with different pitches, or different angles of inclination of the strokes on the scales. A pattern with different pitch values is known as a Vernier pattern.

Simplified expressions for two raster Vernier patterns were developed, as a function of $(i, \Delta\omega)$, where i is the coefficient of optical reduction and $\Delta\omega$ is the difference between the pitches of these two raster scales. The Vernier pattern parameters can be expressed on this basis and are listed as shown below.

Pitch of the rasters (μm or sec. of arc):

for the first scale $\qquad\qquad$ $\omega_1 = i\Delta\omega;$

and for the second scale $\qquad$ $\omega_2 = (i-1)\Delta\omega;$

Period of the Vernier combination $\qquad$ $W = i(i-1)\Delta\omega;$

The number of periods: in the circle $\qquad$ $M = \dfrac{360°}{i(i-1)\Delta\omega};$

and for the linear pattern $\qquad\qquad$ $M = \dfrac{L}{i(i-1)\Delta\omega};$

(*L* is the length of the scale).

The other parameters of the combination can be calculated using the expressions presented above. The fringe patterns of the combination of two linear raster scales are shown in Figure 3.14 (a, Vernier; b, Moiré pattern).

It is possible to calculate the light transmission function Π and the modulating function *m* in the period of the Vernier pattern. These functions are given below by

$$\Pi = \frac{1}{W}\sum_{j=1}^{i-1} a_j = 0,25$$

$$m = \frac{\left[\sum_{i=1}^{n}(i-k) - \sum_{k=1}^{i}(k-1)\right]\dfrac{\Delta\omega}{2}}{\left[\sum_{k=1}^{i}(i-k) + \sum_{k=1}^{i}(k-1)\right]\dfrac{\Delta\omega}{2}} = \frac{\sum_{k=1}^{i}(i-k) - \sum_{k=1}^{i}(k-1)}{\sum_{k=1}^{i}(i-k) + \sum_{k=1}^{i}(k-1)}.$$

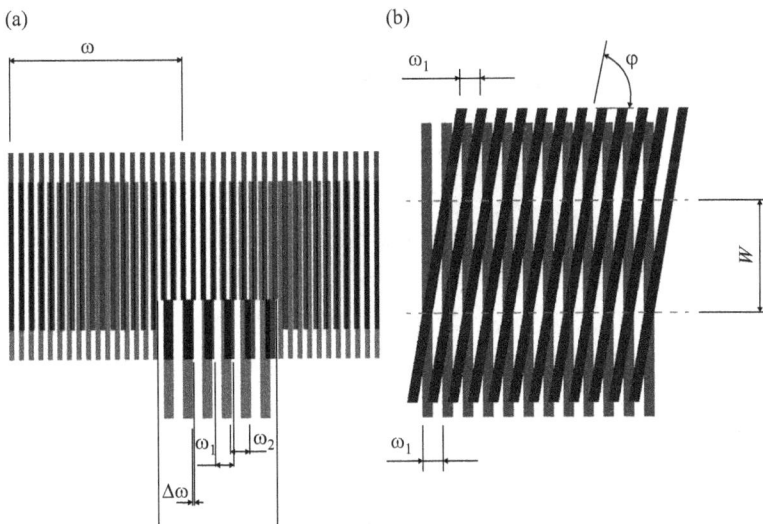

Figure 3.14. (a) Vernier and (b) Moiré fringe patterns of two raster combination.

The light transmission function through the analyzing gap $T_a = k\omega_2$ will be given by

$$\Pi_a = \frac{1}{T_a}\sum_{j=1}^{i-1}\left\{\frac{\Delta\omega}{2}|(i-2j+1)| + \frac{\Delta\omega}{2}\left[\frac{2j}{i+1}\right]\right\}.$$

After some transformation of the equation for m and using the mode of square numbers, the equation reduces to

$$m = \frac{(i-1)+(i-2)+\cdots+0-0-1-\cdots-(i-1)}{(i-1)+(i-2)+\cdots+0+0+1+\cdots+(i-1)} = (i-k)/(i-1).$$

Figure 3.15 shows the dependence of the optical coefficient i, the width of the analyzing gap T_a, and the amplitude A of the light transmission function π in the Vernier combination of the two raster scales.

The equations thus derived, coupled with the results of experiments performed, allow the conclusion that Moiré fringe patterns could be used for the assessment of the accuracy of raster scales. If one of the raster scales is chosen as the reference scale then the other, in conjunction with the first, will show any change from the exact form of the pattern. A displacement of one of the raster scales gives the movement of the period of the pattern multiplied by the value of the optical coefficient i. The same

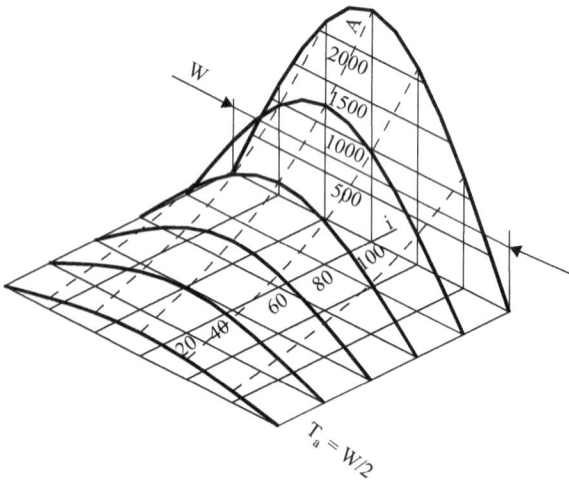

Figure 3.15. The variation of amplitude A of the light transmission function by the Vernier pattern through the different sizes of analyzing gap.

occurs with the error of the raster strokes; it will be magnified by the optical coefficient as well. This optical effect can be used for the determination of raster scale accuracy, taking one as a reference. Scales without errors would form a regular circle for acircular pattern of the scale or a straight line for a linear pattern of the scale. An example of the assessment of the accuracy of the circular scale using a roundness template is shown in Figure 3.16. The method is assumed to apply for control of accuracy assessments of systematic error in the range of 1'...10' (min. of arc) for circular scales and for about 10 µm and more, for linear scales.

The measuring device for the accuracy control of a linear raster scale consists of slide ways of high accuracy, the raster scale to be measured, a carriage with the index (as a master scale) raster and an optical system, consisting of light sources, photocells, and light transfer windows. Graphs can be shown of raster pitch errors representing raster scale accuracy, determined by comparing it with the results received by the usual methods. Diagrams of a device for measuring pitch errors of linear raster scales are presented in Figure 3.17 (a and b).

The main parts of the device (Figure 3.17, a) are the slide ways 7 of high accuracy, 1 the raster scale to be measured, 3 a carriage with the index raster 2 and an optical system consisting of light sources 4 and photocells 5, 6 and light transferring windows. In Figure 3.17(b) there are graphs of raster pitch errors representing raster scale accuracy, determined by comparing it with the master scale (above), and the errors graph, measured by using the Moiré pattern (below).

As is shown in the diagram, the two analyzing gaps with light sources (4) and photocells (5, 6) are located at a distance equal to fringe spacing,

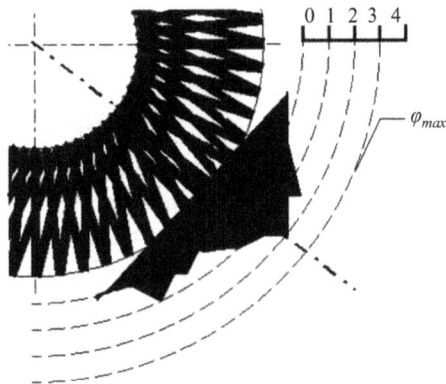

Figure 3.16. Declination from the circle showing the pitch error of the scale.

(a)

(b)

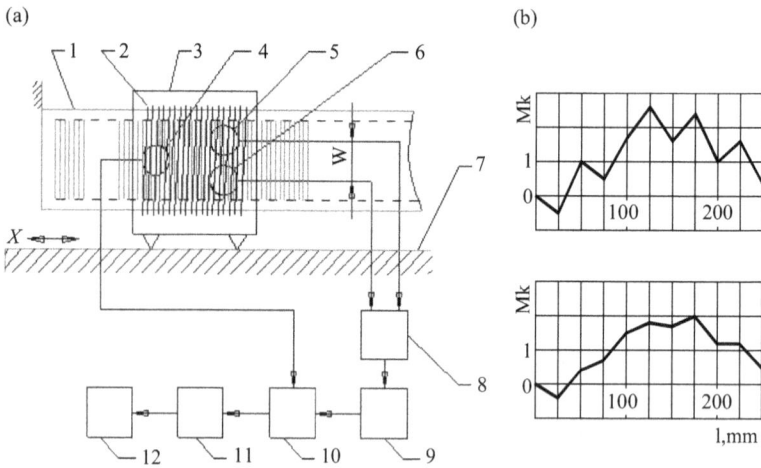

Figure 3.17. Diagrams of measuring the pitch error of linear raster scales: (a) general view of the device, (b) graphs of errors measured.

that is equal to 0.5W or W(k + 0.5); k = 1,2,3.... When the carriage (3) moves along the slide ways (7), the output voltages from the photocells compensate each other. The output voltage from the photocells is amplified and transferred to the summing unit (8) and further to the recording device (9). It is obvious, that in the case of a high-accuracy raster (2), the output voltage in the recording device is without any changes and the graph must be a straight (or sine wave) line. In the occurrence of a raster pitch error, the fringe pattern moves for a distance, equal to $i\Delta l$. The output voltage changes accordingly, providing an evident indication of a raster scale pitch error. The sensitivity of the index device included in the carriage (3) can be tuned enabling a wide range of applications. The measurement can be performed on various kinds of raster scales, using Vernier or other Moiré fringe patterns.

Such a device can be used for measuring the straightness of slide ways, lathe bedding, and so on. In this case the device is used with a calibrated raster scale that has all the other parts as mentioned above. The master straight line is formed between the two points that support the transducer. Any deviation from the straight-line movement of the index head along the slide ways changes the position of the index head resulting in an output voltage from the photocells. The results can be represented in digital or graphical form. A pattern of alternating peaks and valleys on the graph is an indication of some bias in the accuracy of certain strokes of the scale. Graphical results from a Moiré fringe measuring device give a representation of raster scale accuracy, expressed as a result of some stroke groups. It may be referred as the accuracy of the transducer linear displacement where the raster scale is employed. In some way it has more information and is more useful than certain stroke errors detected by comparing it with the master scale or a laser interferometer. There exist simple technical ways to interpret the results in the digital mode and use the facilities of many computer programs for statistical assessment.

Measurement of circular raster scales is accomplished in a similar manner, the only difference being in circular movement of the index device. The raster scale or the index head must rotate on a high-precision spindle because the trajectory of rotation (a circular movement) serves as the etalon of the angular displacement. Precise roundness measuring devices are pertinent for this purpose.

Moiré fringe patterns can be employed effectively for accuracy improvement, and to achieve a higher number of discrete values, in the design of linear and rotary encoders. In the case of rotary encoders the reference circular raster scale with the number of strokes, z_1, is mounted in the housing of the rotary encoder and concentric to it is the measuring raster scale with the number of strokes equal to Z_2. The scanning raster disc has the number of strokes equal to Z_1. This disc is fixed to the bedding by flat springs; one end being fixed to the mounting of the raster scale and the other end is fixed to the bedding. The light source is from a ring-shaped light bulb with alight-transmitting diaphragm. The shaft of the encoder rotates in the bearings together with the disc. The photocells also are ring-shaped and are fixed to the bedding. The signal output is connected to the amplifiers and pulse forming units, and then to the coincidence loop and digital display unit, where the rotary displacement is indicated. The Moiré fringe period formed by those two raster scales moves to an angle equal to one full pitch of the raster scale after the rotation of the shaft for the difference of the raster pitch $\Delta\omega$. The output of the signal is periodic at every pitch of the raster scale. Thus at every rotating angle of discrete value $\Delta\omega$, the digital encoder unit is counting the pulses of rotation. During a full rotation of the shaft without an additional interpolation unit the encoder will output Z_1 Z_2 number of signals, the discrete value of which is $\Delta\omega = 2\pi / z_1 z_2$.

3.4 MEASUREMENT OF LINEAR SCALES AND TRANSDUCERS

IMS are calibrated against the reference standards of measure comparing their accuracy at some pitch of calibration, for example, at the beginning, middle point, and the end of the stroke, or at every tenth stroke, or over the range of measurement in the total length, or the circle (for angular displacement). The calibration depends on the written standards and methodical documentation of the machines or instruments. During the calibration it is possible to find out only a restricted number of values. An example of the results shows that during the accuracy calibration some larger values of the error can be omitted, including significant ones. So, it is important to determine the information quantity on an object that was assessed, providing more complete measurement during the calibration processes.

The intervals of verification of the accuracy of linear transducers have been checked against the reference measure, changing the length of the intervals. The digital output of photoelectric translational and rotary transducers have the last digit number equal to a value of 0.1 μm or 0.1″ (arc). The measuring range in these cases is equal to 10 to 30 m, or a full or several rotations of the shaft to be measured. The value in arc seconds of one revolution is 1,296,000″, which is equivalent to 12,960,000 discrete values in the display unit. However, the measurement results in the display unit can be proved by metrological means only at every 100, 1,000, or a similar number of strokes. Information entropy allows that part of the information available from the measuring device that is assessed by metrological calibration means to be presented to the user. So, the analysis presented here is concerned with the information evaluation of the measuring system of the machines. Practically all moving parts of the machine together with its IMSs take part in the manufacturing of a part on this machine. Therefore, all inaccuracies of the IMS translate into inaccuracies of the part produced. Information entropy provides a vehicle to reach the cognitive information about the accuracy available in the system under consideration. A general diagram of the sequence of production, control, and implementation of raster scales into transducers and the IMSs of machines is shown in Figure 3.18.

3.5 COMPARATORS FOR LINEAR MEASUREMENTS

The linear displacement errors of linear displacement transducers are determined by using an automatic comparator. A high efficiency of measurements (lasting about 10 minutes for each measurement together with the presentation of results) is required to achieve a sufficient quantity of data for a statistical evaluation. This provides the basis to perform the next step in the accuracy improvement of machines, that is, to construct

(a)

(b)

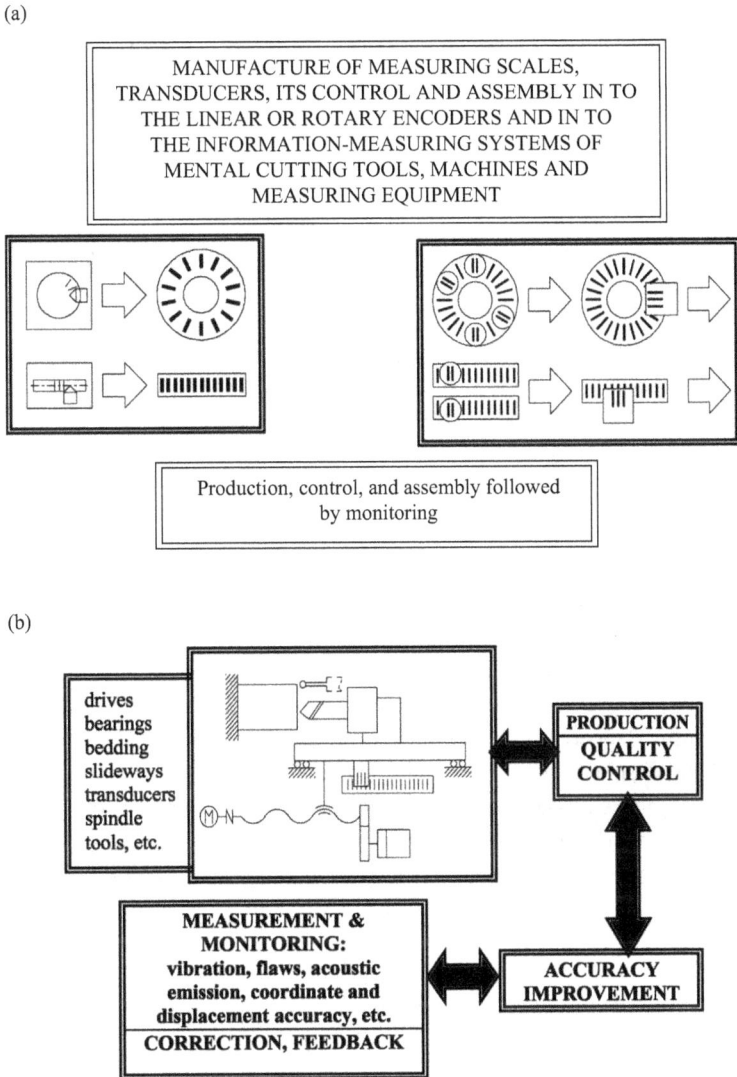

Figure 3.18. General diagram of elements of a machine where precision operation is to be controlled and monitored: (a) elements of the information-measuring system, (b) machine's parameters for measurement and monitoring at the industry level.

the diagrams and methods for error compensation. The calibration intervals were evaluated by statistics, and a histogram of the comparison of the results was developed. The automatic lengths and transducer comparators created can be regarded as the practical means for the calibration of IMSs.

The design and purpose of a comparator for measurement and calibration of linear raster scales and translation transducers is as follows:

- Translation transducer compared with laser interferometer as the standard measure.
- Translation transducer compared with the other translation transducer that serves as a reference measure.
- One linear raster scale, reading the positions of the strokes by PMs, enabling the measurement of 0.005—0.1 mm width of the strokes, is compared with the laser interferometer.
- Linear raster scale with PMs is compared with the other linear raster scale as a reference measure.
- Compare translation transducer with the raster scale and vice versa.

A general diagram of a comparator for linear standards calibration is shown in Figure 3.19. The main parts of the comparator are: (1) base on which the moving table (2) is placed on slide ways of high accuracy (sometimes, on aerostatic bearings); (3) drive; (4) the linear scale to be measured, (5) linear scale used as the reference (etalon) scale; (6 and 7) PMs used with both scales as stroke reading instruments. Outputs from the microscopes are input into the control computer via an interface. The measuring data are processed in the computer (8), systematic and random errors and uncertainty of measurement are determined and the graph of errors is plotted using a printer/plotter (9). The placement of the length measure to be calibrated and etalon (standard) measure must comply with the Abbé principle, for example, to be placed as close as possible in parallel placement or must be placed on one line at the longitudinal axis

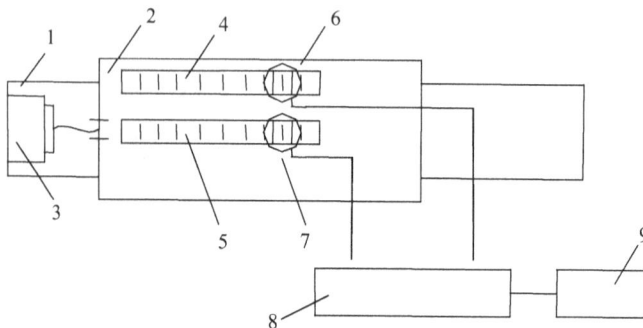

Figure 3.19. General diagram of the comparator for linear standards calibration.

of these measures. Additional means (supplementary interferometer) are sometimes used to eliminate an Abbé offset.

A comparator using the laser interferometer as the length standard is shown in Figure 3.20. The main parts are: (1) base, (2) moving table, (3) drive, (4) the scale to be measured (or translational transducer), (5) PM, (6) mirror for reflecting the laser beam, (7) laser interferometer, adjusted such that the laser beam goes along the symmetry axis of the scale or transducer to be measured, (8) control computer, and (9) printer/plotter of the protocol or graph of measurement.

During the displacement of the table, signals from the PM are compared against the ones from the laser interferometer at the predetermined pitch of measurement. The difference means an error of measure in control. Modern laser interferometers perform correctional assessment for temperature, humidity, pressure changes during the measurement, and for the differences in temperature expansion coefficients of the materials of the etalon and controlled objects.

An important feature of the comparator is the dynamic mode of operation. Measurement is performed during the movement of the scale or the transducer to be measured. Such measurement shows the errors in working conditions of the scale or transducer, including the influence of dynamics of the machine. The enhanced velocity of the table movement during the interval between the measurement points is about 0.2 m/min, and during the measurement of the stroke position the table moves at 100 μm/s. The working cycle of the comparator consists of preparatory operations, where the number of measuring reciprocal strokes, the pitch of measurement, and other parameters are preset in the computer. A wide range of commands enables measurement with many varieties of pitch, changing the beginning of the scale's position, including correctional or compensation parameters for temperature, pressure, humidity and material differences, and so on.

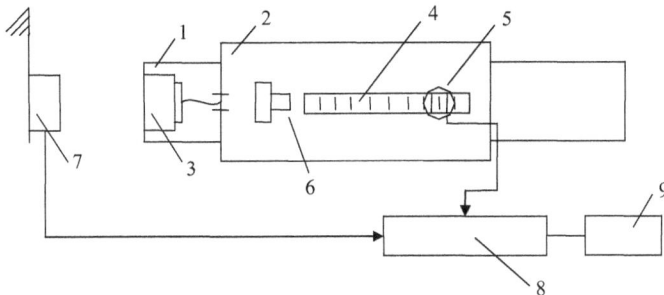

Figure 3.20. The comparator with the laser interferometer as the standard of measure.

The measurement environment is kept within very stringent conditions, including an air conditioning system, temperature regulation, and damping of the base of the comparator against shock and vibrations. The results of measurement can be presented in the form of a printout protocol, an error graphic diagram or just the maximum error results. The arithmetic mean value of the measurement results is calculated, the mean square value, the dispersion, and the maximal positive and negative values of errors are determined. A diagram of systematic error can also be presented if required. In the case of serial measurements, the accuracy of measurement is lower because the efficiency or productivity of measurement is more important. Many individual adjustments and preparation operations are neglected and the cycle of measurement is simplified using special fixtures.

A more detailed diagram of the comparator is shown in Figure 3.21. The main parts of the comparator are: (1) base, (2) table, (3,4) reference measure and scale or LTT to be measured, (5,6) PMs, (7–13,17,18) drive and electronic units, (14–16) computer, (19–21) laser interferometer, and (A–K) analog code converter.

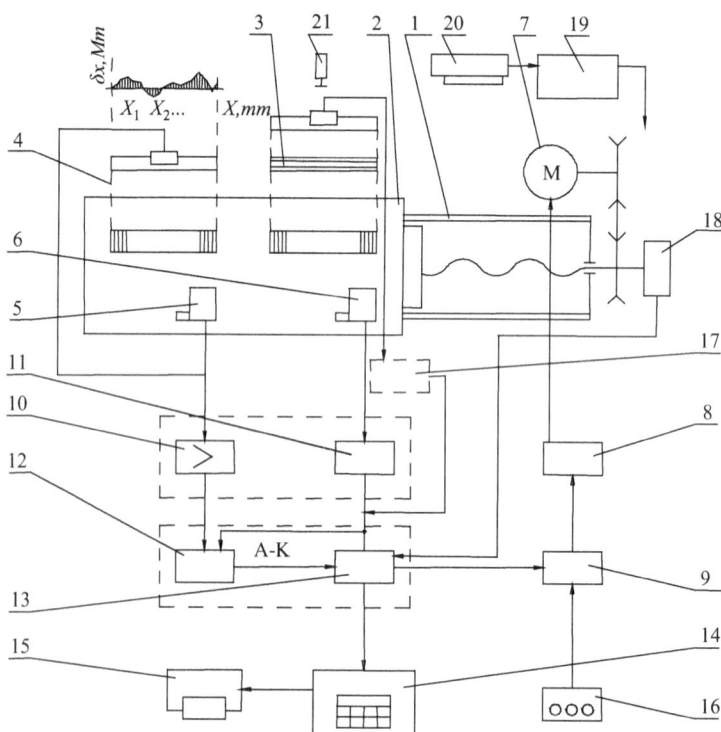

Figure 3.21. A more detailed diagram of the comparator.

Figure 3.22. The casing of the comparator for length standards measurement.

In the case when there is a need to protect the comparator from environmental influences a casing is applied as shown in Figure 3.22. The casing helps to maintain a stable temperature, to protect from dusty environment, humidity, and so on. It is especially helpful when the measurements are not performed in a laboratory.

3.6 COMPARATORS FOR ANGLE MEASUREMENTS

Precision raster and optical scales, together with rotary encoders, are the most common angle measures in relevant instruments for angle measurements. Precise angle encoders are widely implemented in modern industrial and other machines. Such encoders are also used in many geodetic instruments such as electronic theodolites, total stations, laser-scanners, and so on. Calibration of such devices is quite complicated since the encoders used generate a large number of angular values which cannot be tested using classical means of angle determination such as multiangular prisms with autocollimators, and so on, which can be applied for testing only a limited number of angular positions. To try to accomplish such tasks a test bench for testing and calibration of angle-measuring geodetic instruments has been designed and developed. The classical principle of comparison is implemented for angle calibration purposes, which allows itto be used not only for testing geodetic instruments but also for many other angle-measuring instruments, such as rotary encoders, circular scales, and so on.

The effectiveness of the application of piezoelectric actuators is proved by the construction of a comparator for angular measurements.

The comparator for the calibration of circular standards (Figure 3.23) consists of the housing, rotary encoder as a reference measure of angle, elastic clutch, rotary table, PMs, the reference scale to be measured, piezoelectric drive, basic parts for the scale or transducer mounting, and the fixing to the rotary table. The reference scale is used as a complementary reference measure with PMs in the case when the reference measure, the rotary encoder, is not appropriate for the calibration process.

The comparator is mounted in the housing with a high-accuracy aerostatic spindle together with the rotary encoder and the scale as a reference measure of angle. The scale or another rotary encoder to be measured is adjusted concentrically and by a leveling device on the rotary table. One of the main requirements for such type of equipment is the accuracy of rotation of the table on the axis. Applying piezoelectric plates, fixed by elastic suspensions to the rotary table, efficiently solved this task. Symmetrical positioning of the plates and elliptical movement of the contact point to the table permits the transfer of movement, that is, rotation of the table with the highest accuracy, not distorted by the driving mechanism. Experiments undertaken confirmed the high accuracy of rotation of the table using piezoelectric actuators and the run-out of the axis under rotation was within 0.1 μm. Other types of mechanical rotation drives have much more influence on the accuracy of the table rotation. This parameter is important for circular scales measurement as every axis run-out has an influence on the readings from the strokes of the scale. As mentioned previously, in the case of serial measurements or accuracy monitoring, the accuracy of measurement is lower because the efficiency or productivity of measurement is more important. Some individual adjustment and preparation operations can be omitted for monitoring purposes and the cycle of measurement is then simplified.

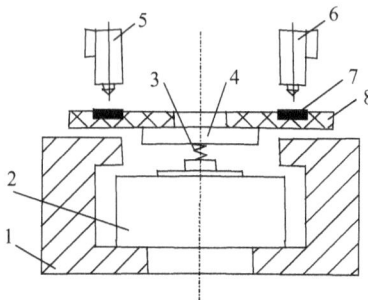

Figure 3.23. The main principle of the comparator for angle calibration.

A diagram of the angle comparator is shown in Figure 3.23. The main parts of the comparator are: (1) housing, (2) rotary encoder of high accuracy as the standard of measure, (3) coupling, (4) rotary table, (5,6) PMs, (7) circular scale to be measured, and (8) strokes of the scale. The rotary encoder 2 is connected with the rotary table via coupling 3 and table 4 is mounted concentrically. Table 4 is equipped with an adjustment device for adjustment of the circular scale 8 in the radial direction and for leveling its flat surface in the horizontal direction, perpendicular to the axis of rotation. The PMs are positioned above the strokes of scale 7. Microscopes 5 and 6 are placed on diametrically opposite sides with the purpose of eliminating any eccentricity of the scale during measurement.

During the measurement the readings from the microscopes are summed and divided by 2, $\left(\dfrac{M_1 + M_2}{2} \right)$, where M_1 and M_2 are the readings from the PMs, thus eliminating any error from an eccentricity. The drive is not shown in the diagram. As a reference measure, the etalon of measurement, either a high-precision rotary encoder or a circular scale of high precision can be used. They must be calibrated against a higher level angle-measuring etalon. Usually, one of the classical methods of calibration are used (Jelisiejev, Heuvelink, Wild, Bruns, etc.) but the 1440 Index Table or phase shift methods are also used.

Figure 3.24 shows a prototype angle calibration test bench. The runout of the air bearing was measured using a laser displacement transducer

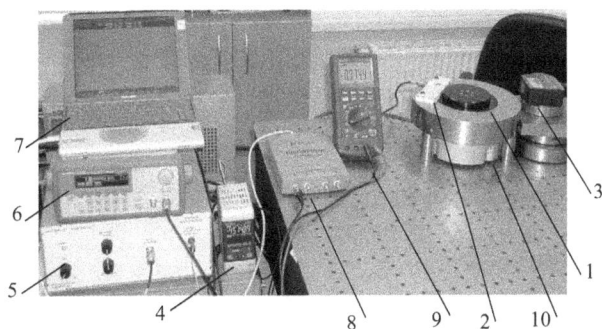

Figure 3.24. Prototype of angle calibration test bench.
(1) aerostatic bearing RT100L Nelson Air Corp, (2) piezoactuator (drive), (3) laser transducer of small displacements LK-G82, (4) control unit of the transducer LK-GD500, (5) mains amplifier EPA-104, (6) signal generator Agilent 33220A, (7) personal computer, (8) analog-digital converter PICO 3424, (9) multimeter, and (10) precision rotary encoder Heidenhain GmbH.

LK-G82. The piezoactuator was connected to the 18 V voltage and a 21.3 kHz harmonic signal was supplied.

The piezoelectric plates are supplied by high-frequency voltage oscillations from the generator. The mechatronic control of the actuator allows the velocity of rotation of the table to be changed within wide limits. Very slow adjustment movements are also available. They are applied during the setup of all measuring devices of the comparator. The PC controls by mechatronic means the comparator for the circular scales and rotary encoders calibration as the data received during the calibration is very extensive. The feedback for the comparator's control is received from the reference encoder's signals to transfer data from the PMs. The data from the microscopes are compared with the readings from the reference rotary encoder for the final calibration data evaluation. The operational cycle of the comparator consists of:

Preparation cycle.
Adjusting and leveling of measuring objects. Adjustment of output signals.
Selection of the mode of measurement.
Stabilization of temperature and pressure.
Selection and input of correctional factors.
Automated measurements.

A multipurpose test bench for calibration of angle values using several different angle standards of measure is shown in the diagram in Figures 3.25 and 3.26.

The measurement process can be performed with periodic stops and also during continuous rotation of the spindle with the reference measure and the object to be measured. Some technical specifications of the comparator are presented below:

Velocity of rotation of the spindle, rev/min	0–10.
Maximum spindle rotation trajectory run-out: in radial and axial directions, µm	0.1.
Eccentricity of raster scales, not exceeding, µm	0.5–0.8.
Diameter of the scales: maximum, mm	400.
minimum, mm	160.
Number of signals of rotary encoder to be calibrated, p/rev	100–32,400.
Minimum value of signal of reference measure (")	0.1.
Form of presentation of the measurement results:	graphical; protocol of the results.

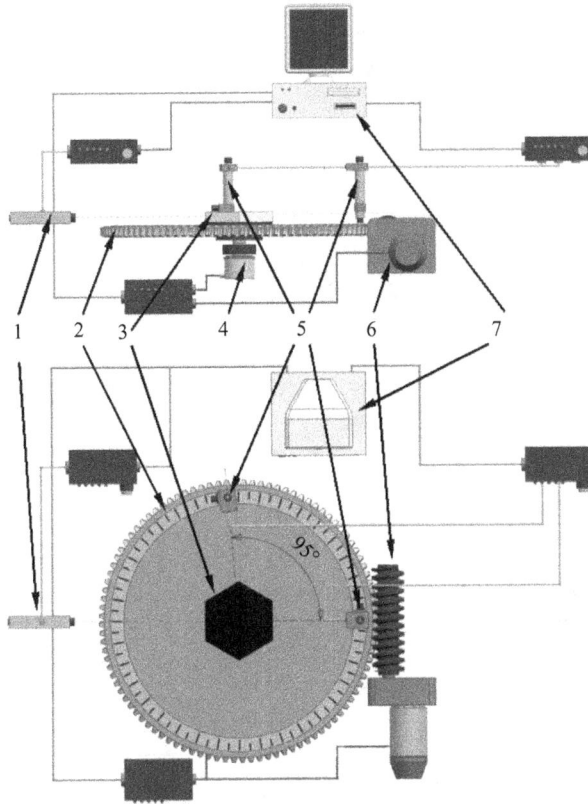

Figure 3.25. Layout of the angle-measuring device: (1) auto-collimator, (2) rotary table with the circular scale, (3) multian-gular prism, (4) photoelectric angle encoder, (5) microscopes, (6) worm gear drive with step motor, and (7) control PC.

Using a circular scale as a reference standard, a multivalue device for an easily conveyed angle calibration of high discreteness and accuracy, or even a flat angle calibration standard, can be created. The optimum structure of a flat angle calibration bench would possess the technological characteristics indicated in Table 3.2.

The unique feature of the test bench developed (Figure 3.26) is the implementation of three independent principles for angular position deter-mination within the same equipment, with the possibility of using these principles simultaneously. The three means of angle determination are the photoelectric rotary encoder, the circular scale (1) with microscope(s) (6), and the multiangular prism (2) with the autocollimator(s) (3). The entire equipment, including the angle position reading optical instruments like

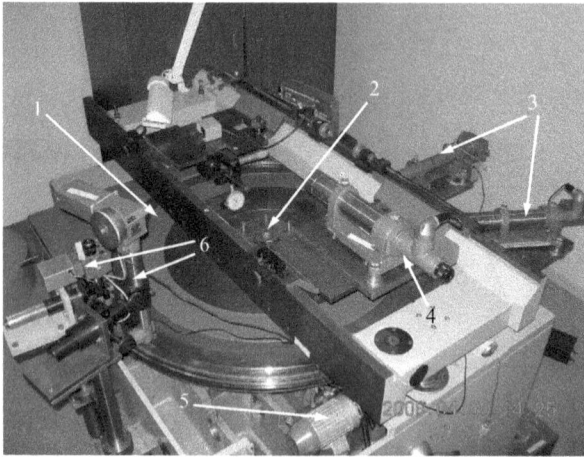

Figure 3.26. Layout of the measuring equipment: (1) rotary table with the circular scale, (2) multiangular prism, (3) autocollimators, (4) autocollimator for determination of the position of the tested instrument, (5) worm gear drive with step-motor, and (6) microscopes.

Table 3.2. Accuracy and technical parameters of a test bench for flat angle calibration

No.	Structural elements	Parameters
1.	Hardware	Rotary table with the axis, circular scale, microscopes, autocollimator and polygon
2.	Total run-out of axis, μm	0.05–0.1
3.	Number of scale strokes	1080 ($\varphi_t = 20'$)
4.	Width of the strokes, μm	5
5.	Microscopes	Photoelectric
6.	Standard deviation of the microscopes, S, μm	0.05
7.	Multiangle polygon	12–72 angles
8.	Accuracy parameters of the angles of the polygon	$S = 0.03$; $P = 0.99$
9.	Autocollimator	Measurement range $\pm 10'$; $S = 0.02$; $P = 0.99$
10.	Pitch of angular positioning of the rotary table, degrees	$1'$; $10'$; $20'$; $1°$; $3°$; $10°$

microscopes and autocollimator, is controlled by the PC which transmits a command for the table rotation via the step motor and worm-gear (5). For determination of the small angular fluctuations of the geodetic instrument under testing (tacheometer, theodolite, etc.), a special autocollimator (4) has been fixed. The machine has a thermally stable base where high-precision rolling bearings are mounted, onto which is fitted the rotary disk and circular scale. The machine was powered electromechanically by means of a worm gear with agear-ratio 1/1080. A high-precision circular scale dividing machine was applied as a base for this equipment.

To ensure the efficiency of the test bench the measuring instruments were directly connected to the computer so that readouts could be taken without the intervention of the operator into the measurement process (thus decreasing potential operator errors). Hence, several optical instruments (namely microscopes and autocollimators) were modified by fitting CCD matrixes inside them (Figures 3.27 and 3.28). Since all the measurements are performed in a static mode, low-speed CDD matrixes are used.

Using a CCD matrix instead of a CCD linear light detection sensor reduces the influence of optical system distortion of the instrument and (in the case of the autocollimator) reduces the errors caused by the flatness deviations of the reflective mirror (or multiangular prism face). The CCD matrices also significantly increase the accuracy and sensitivity of the measurements taken by these instruments.

Using the CCD matrices reduces the standard deviation of measurements to $0.0125''$ in the case of the microscope and $0.041''$ in the case of

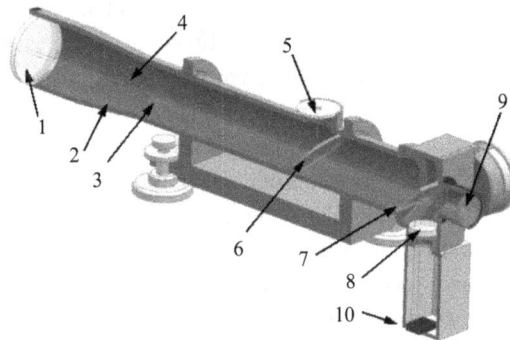

Figure 3.27. Modernized autocollimator, cut section: (1) objective lens, (2) autocollimator body, (3,4) sent and returning light beam, (5) the light source, (6,7) light reflecting/transmitting plates, (8) lens, (9) ocular, and (10) CCD matrix.

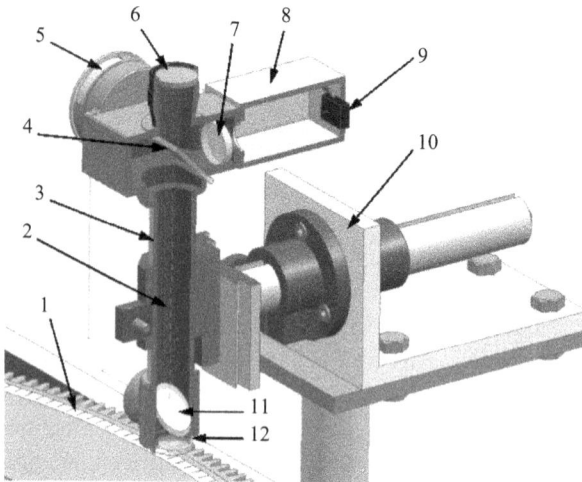

Figure 3.28. Cut section view of the modernized microscope:
(1) circular scale on the rotary table, (2) optical axis,
(3) microscope body, (4,11) light reflecting/transmitting plates,
(5) measuring drum, (6) eyepiece, (7) lens, (8) additional box,
(9) CCD matrix, (10) microscope holder, and (12) objective.

the autocollimator measurements; the resolution also increased to approx. 0.002″ for the microscope and approx. 0.001″ for the autocollimator.

Therefore, it may be stated that the circular scale microscope method of angular position measurement can be successfully implemented with a precision close to that obtained by the multiangular prism and autocollimator but at a much smaller pitch. The small pitch of measurement is especially important for testing or calibration of angle-measuring geodetic instruments as well as for any other type of modern instrument for angle measurements (angle encoders, raster or coded circular scales, etc.).

The test bench developed incorporates several precise angle-measuring techniques and reference measures of angle (photo electrical angle encoder, polygon autocollimator, circular scale microscope). This allows independent and cross-calibration measurements. Optical instruments such as autocollimators and microscopes used in the test bench were modernized by fitting them with custom-made CCD matrices and by developing special software. The new scale calibration method based on constant angle setting in a full circle with multiple turns was tested on the plane angle testing/calibration bench. The circular scale was calibrated using two PMs placed at an angle of 95° to each other with a pitch of calibration equal to 5° with additional control using the multiangular prism autocollimator.

3.7 CALIBRATION OF GEODETIC INSTRUMENTS

In geodesy, surveying, machine engineering, and other branches of industry there are very widely used instruments that allow precise planar angle measurements. Such instruments are theodolites, digital theodolites, total stations, and so on. In common with all other measuring instruments these instruments must be tested and calibrated and this is regulated according to ISO 17123-3 [24] and ISO 17123-5 [25]. According to the standards the accuracy of the angle measurement performed by the instrument must be tested in field conditions using the known length reference measure for the angle measurement (triangulation principle). Using such a method it is possible to obtain only a very restricted number of angular measurements. It doesn't allow for the collection of a large number of different (desired) tested angular values. On the other hand tested geodetic instruments display a vast number of discrete values on their display unit during measurement, and these values must also be checked.

The first of the methods presented is based on the precise multiangular prism (polygon). Normally this has between 12 and 72 flat mirrors positioned at a very precise constant angle to each other. The polygon usually is turned to a certain position together with the object to be measured and the angle of rotation is registered by the optical instrument, the autocollimator. Such a method is a very widely applied measuring technique in geodesy instrumentation as well. This method has one shortcoming, that is, the discretion of this method is large, and hence it is only possible to check a small number of values offered by the calibrating measuring instrument.

The second method presented (see Table 3.1) uses a very precise tool, Moore's 1440 Precision Index. Moore's 1440 Precision Index is an angular measuring device consisting of two serrated plates joined together to create the angle standard of measure. During measurement the upper disk of the Index is lifted, the lower part rotates with the object to be measured, after that the upper part is lowered back and the readings are taken. The method has a high precision of 0.004″ (Table 3.1), although it also has some shortcomings. It is very difficult to automate, also during lifting of the table (which is a necessary technological operation of the method) the calibrated instrument may become less stable, move, and unexpected errors may occur.

The third method presented is a classical one both in geodesy and in general engineering. This method has been very widely used in the past and it requires a highly accurate circular scale and one or more (depending on the measuring method) microscopes (preferably photoelectric) for the scale readings. The major shortcoming of this method is the need of a circular scale of very high accuracy; the scale must be of a large diameter for positioning the PMs, and it requires precise manufacturing and calibration is time consuming. Due to the high cost of such processes this method is gradually being replaced by rotary encoders.

The fourth method presented is the most widely used nowadays; it uses digital rotary encoders as the reference measure. Using a modern high-accuracy digital rotary encoder it is possible to achieve a very good result comparable with the classic methods. Using rotary encoders also allows a reduction in the size of the test bench and the system can be easily automated.

To use the angle-measuring methods discussed above for the calibration and testing of geodetic instruments, their accuracy must be higher than the accuracy of the instruments being calibrated. Standard deviations of the horizontal angle measurement of the most commonly used electronic tacheometers are listed in Table 3.3. Consideringthese technical specifications it can be considered that all the angle-measuring methods listed in Table 3.1 could be used for their calibration and testing, the difference being in whether they are more or less suitable for the particular task.

Some typical technical parameters of different types of geodetic instruments are as follows:

- values of circular scale's division: 1°; 20′;
- scale reading values of a microscope of an instrument: 5′; 1′;
- accuracy of readings from a circular scale: 0.5′; 0.1′; 15″; 1″; 0.2″;
- standard deviation of angle measurement   1.5″ 0.5″.

They show a high accuracy of measurement by using precise instruments, so a higher level of accuracy calibration of such instruments is needed. The accuracy data presented above forces one to apply a measurement standard for calibration purposes of higher accuracy than the accuracy of the instrument itself.

The calibration of geodetic-measuring instruments requires a large number of angular values to be compared with the reference values. Such a procedure, due to its technical complexity, is not regulated by any European standard and very few test rigs are available to perform the complex testing and calibration of planar angle for geodetic instruments. These devices are usually operated by the manufacturers of the measurement equipment and are not available for the wider public and the users of these

Table 3.3. Technical specifications of the most commonly used electronic tacheometers

Instrument model	Standard deviation of the angle measurement	Instrument model	Standard deviation of the angle measurement
Leica		**Trimble**	
Leica TC®707	7″	Trimble 3303; 5603	3″
Leica TC605; TC®705	5″	Trimble 3305/6	5″
Leica TC805; TC®703	3″	Trimble 5601	1″
Leica TPS1201	1″	Trimble 5602; 5602 DR300	2″
Leica TCA1800	1″	Trimble S6	1″
Leica TCA2003	0,5″		
Sokkia		**Topcon**	
Sokkia SET500; SET4C/SET4B; Sokkia SET5F	5″	Topcon GTS6A/6B	2″ – 5″
Sokkia SET600	6″	Topcon GTS1	20″
Sokkia SET2C/SET2B	2″		
Sokkia SET3C/SET3B	3″	*Hewlett—Packard*	
Sokkia SET6F	7″	3820A Elect. Total St.	2″

instruments. It is obvious that there is a need for a test bench to be developed, which is capable of performing the plane angle testing and calibration of geodetic instruments.

A worm gear drive is used for transporting the object to be measured into the required position. The polygon and autocollimator, the rotary encoder or the circular scale with microscopes can be chosen as the standard measure for angular displacement control. When using a circular scale as the standard of measure, two PMs are used to avoid the influence of eccentricity for angular measurements.

3.7.1. METHODS AND MEANS FOR VERTICAL ANGLE CALIBRATION

A test rig developed for accuracy monitoring of geodetic instruments is shown in Figure 3.29. All the items shown in the picture are mounted on the sturdy base of the test rig. The parts of this equipment are identified as follows. The worm wheel WD is driven by the step motor SM and on the surface of the worm gear there is a circular scale to which the PMs M1, M2, and M3 are pointed. The light beam from the autocollimator AC is directed via the display unit DU onto the side of the multiangle prism (polygon). The measuring information from the AC is input into the computer PC. For automating the monitoring process the rotary encoder RE is coaxially fixed to the worm gear, the output from which is fed via the control unit CU into the PC and to the step motor SM. Also the geodetic instrument GI is fixed to the axis of the rig with the possibility of axial adjustment and leveling. The GI is pointed to the target T1 for horizontal angle measurement and to the target T2 for vertical angle measurement.

The mode of operation of the test rig is as follows. Using the worm gear and the step motor controlled by the computer, the rotary table rotates to a certain desired angle measured by the rotary encoder and stops. The table can be positioned with a discretion of $0.001°$ and the angle position is measured by PMs using the circular scale. The photoelectric autocollimator is pointed at the multiangular prism (12 angles) and the discretion of these measurements is $30°$.

The geodetic instrument (electronic tacheometer) is attached to the rotary table by its base and so it rotates together with the table. The upper

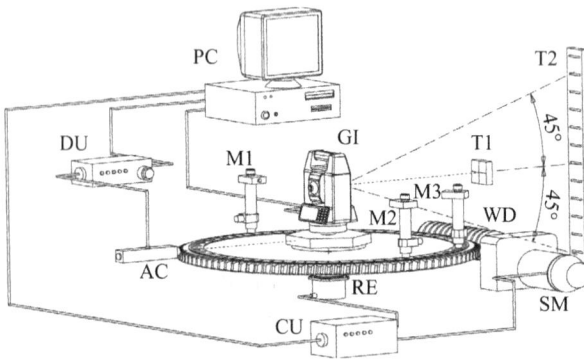

Figure 3.29. A test rig for accuracy monitoring of geodetic instruments against angle standards, including the vertical angle.

part of the tacheometer is loosely supported to the base of the test rig and so it stands still. Since it is very difficult to hold the upper part in an absolutely constant position, to achieve good measuring results the tacheometer is manually pointed to a certain mark at every measuring stop.

The measuring methods using a rotary encoder, autocollimator and polygon, circular scale and microscopes, combined in one test rig can perform the measurements both independently and within a united system. Such a combination of measurement methods allows constant monitoring of the performance of each measuring device and the self-calibration of each component of the system. The multiangular prism and autocollimator system are considered as the reference angle and all the other measuring systems are compared to it. The system makes a full calibration turn (360°) in approx. 30 minutes.

Generally there are several groups of plane angle measurement principles (methods):

1. Solid angular gauge method:
   • polygons (multiangular prisms);
   • angular prisms;
   • angle gauges, and so on.
2. Trigonometric method (angle determination by means of linear measurements).
3. Goniometric method (plane angle determination by means of a circular scale):
   • full circle (limb, circular code scales, etc.);
   • nonfull circle (sector scales).

Calibration and testing of geodetic angle-measuring instruments has always been a serious problem and although calibration of the horizontal angle measurements could be quite efficiently accomplished using standard precise turn tables, calibration of vertical angle measures required some special instrumentation. The calibration was usually performed by a special bench consisting of autocollimators attached at different vertical angles to the calibrated instrument [26] as shown in Figure 3.30(a). Hence the entire test bench was extremely bulky and was able to measure only a very limited number of vertical angles. A new approach to the problem was the implementation of the precise angle encoder for the creation of a vertical angle reference (Figure 3.30, b). In this case it was possible to create an unlimited number of reference angle values, but the equipment was extremely expensive.

Most geodetic instruments have two angle reading devices installed, for horizontal and vertical angle measurements. A number of methods

(a)   (b)

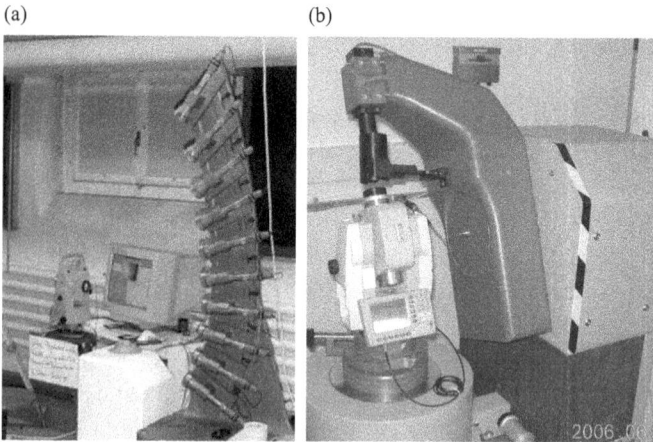

Figure 3.30. Vertical angle calibration of geodetic angle-measuring instruments: (a) implementing a number of autocollimators positioned at certain angle, and (b) implementing the precise angle encoder.

of calibration of the horizontal angle measurements are implemented in practice, based on calibration of circular scales and rotary encoders. An arrangement is shown below to create a reference standard for vertical angle calibration purposes in a laboratory environment. The principle of the proposed vertical angle calibration method is based on trigonometric angle determination (angle determination by means of linear measurements) and is shown in Figure 3.31.

The arrangement uses a reference length scale for vertical readings of the tacheometer and another reference measure of length, for the determination of the distance from the tacheometer's axis to the vertical scale. The designations in Figure 3.31 are: I^1—initial instrument's position with the axis of rotation of the spyglass O_1; I^2—auxiliary instrument's position with the axis of rotation of the spyglass O_2. These positions are achieved by moving the instrument along the slide ways of the test bench used for geodetic instruments testing. At the distance l from the axis of the instrument the linear scale S is fixed perpendicular (vertical) to the instrument's horizontal axis. The distance between both positions of the instrument l_e is fixed by using a reference length measure, for example an end length gauge (length standard). A linear photoelectrical transducer, laser interferometer, or even a precise linear optical scale (with a microscope) can be used for the determination of the distance from the axis of the instrument to the surface of the scale, as it is quite a complicated task to do that initially. For precise vertical angle measurements the distance from the

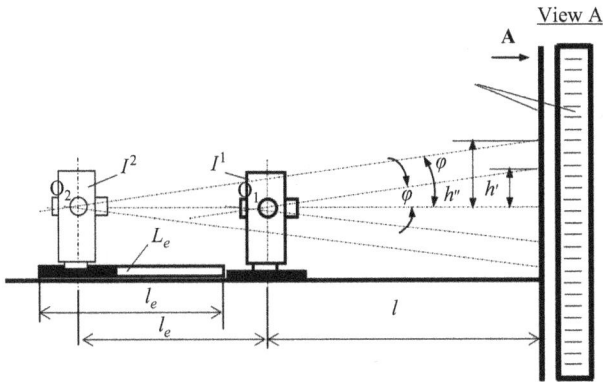

Figure 3.31. Arrangement for vertical angle calibration of geodetic instruments.

instrument to be calibrated (tacheometer) and the reference measure (linear scale) l has to be determined quite precisely (down to 0.01 mm). The accuracy of distance determination influences the results of measurements considerably, giving a bias of the reference data. At the position I^1 of the instrument the reading h' from the scale is taken at the angle φ between the axis of the telescope of the instrument and the horizontal line.

The angle of interest is expressed as

$$\varphi = arctg \frac{h'}{l}.$$

After displacement of the instrument linearly to the subsidiary position I^2, keeping the same vertical angle φ, the next reading h'' is taken and the distance $l + l_e$ can be determined

$$\varphi = arctg \left(\frac{h''}{l + l_e} \right)$$

and substituting the equations, one in the other, will yield the result

$$l = \frac{h' l_e}{h'' - h'} \quad \text{or} \quad l = \frac{l_e}{\dfrac{h''}{h''} - 1}.$$

After taking the readings h' and h'' from scale S, the true value of the distance l will be determined. Further measurements can be performed determining every tested vertical angle of the instrument operating with a known distance and using the readings h_i from the scale. A full range of vertical angles of the geodetic instrument can be tested in the laboratory environment, in such a way improving the accuracy of calibration. Therefore, the real distance from the calibrated instrument to the reference scale can be determined using only a linear scale.

A calibration test of vertical angle measurements of a geodetic angle-measuring instrument (tacheometer) was performed at the Institute of Geodesy, Vilnius Gediminas Technical University. The calibrated instrument was a *Trimble 5503* tacheometer (having a stated standard deviation of angular measurements of 5 arc sec).The experiment was arranged according to Figure 3.31; the tacheometer was mounted on a linearly sliding plate and aimed at a linear scale positioned vertically at a distance of ~2.5 m from the tacheometer. The industrial laboratory high-accuracy linear scale had a length of 1 m and scale strokes at every 1 mm. The linear movement of the calibrated tacheometer was tracked using a laser distance meter and a mirror prism mounted on the sliding plate (together with the tacheometer). After calculating the linear distance from the tacheometer to the scale according to Figure 3.31, it was determined that the measured distance was equal to 2.4215 m. The main objective of the experiment was to test the calibration method and obtain preliminary results of the systematic errors (biases) of the vertical angle measurements using the *Trimble 5503* tacheometer.

The calibration was performed using a 1 m precise linear scale, collimating the spyglass to the scale strokes at a pitch of 10 and 12 mm. The resulting calculated deviations of the tacheometer readings are shown in Figures 3.32 (10 mm pitch) and 3.33 (12 mm pitch).

The standard deviation of the collimation (pointing to the scale stroke) was evaluated as 2.15″ and the general standard deviation evaluation of

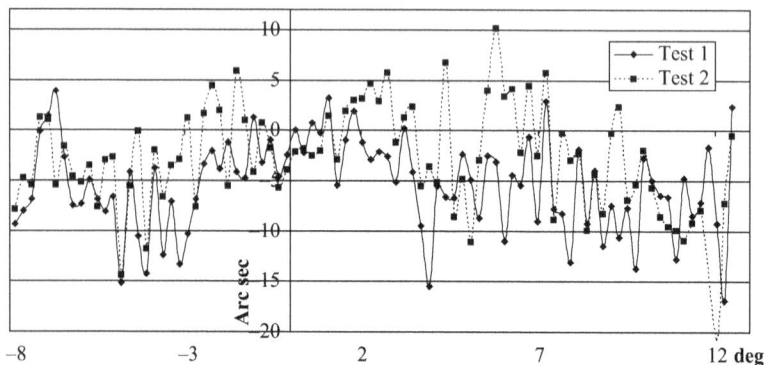

Figure 3.32. Deviations of vertical angle measures determined at the linear pitch of 10 mm.

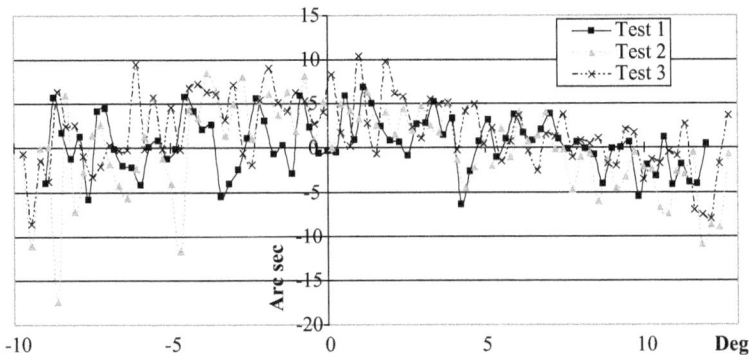

Figure 3.33. Deviations of vertical angle measures determined at the linear pitch of 12 mm.

the entire measurement procedure was 2.31″. As can be seen from Figures 3.32 and 3.33, there is no noticeable short period systematic constituent in the deviations (although more measurements should be performed to confirm this). There is a tendency toward a decrease of accuracy around the limits of the measurements (−10° and 12°), which is quite common for tacheometers, due to both the errors of collimation at steep angles and the principle of action of the vertical angle encoder of the tacheometer (although more tests should be performed to confirm this as well).

Every point at ith angular position accuracy of the instrument is calculated as

$$\delta\varphi_{ti} = \frac{1}{n}\left[\sum_{i=1}^{n}\left(\Delta\varphi_{t_ji} - \Delta\varphi_{t_j0}\right) - \sum_{i=1}^{n}\left(\Delta\varphi_{a_ji} - \Delta\varphi_{a_j0}\right)\right];$$

where

$\Delta\varphi_{t_ji}$—tacheometer readings,

$\Delta\varphi_{a_ji}$—autocollimator polygon readings;

j—number of times of the full circle measurement cycles performed; it usually is equal to 3 or 5.

The readings from the microscopescale are calculated in the same way, thus enabling it to be used as a reference measure for further geodetic instrument calibration.

Some additional difficulties occur when pointing the tacheometer to a certain target. Until now it was attempted to point the tacheometer manually but, as can be seen from the results, manual pointing does not give an appropriate precision. Also the participation of the operator in the machine work considerably slows down the calibration process. It is planned to construct an automatic pointing device which would consist of a photoelectric autocollimator fixed on the nonmoving part of the test rig and a mirror mounted on the tacheometer. The tacheometer would be roughly pointed to a target (as is performed now) and the precise position of the geodetic device would be obtained by comparing the tacheometer readings with the autocollimator data. The calibration process would not be interrupted (data from the autocollimator derived automatically) and the method would enable the elimination of the tacheometer mounting eccentricity bias.

Hence, using the test rig developed, the angle values may be measured by three independent measuring systems as the standard measure of angle, the precise rotary encoder, the circular scale and PM, and the polygon with autocollimator, the electronic tacheometer being mounted on the top of the table. The test rig will run in automatic mode and further automation of the testing and calibration process is possible.

CHAPTER 4

CONTROL OF NANO-DISPLACEMENT AND POSITION

The importance of accuracy control of the position and displacement of machine parts is discussed. A variety of touch probes for measuring the surface of the part are described including multidirectional measuring heads used with coordinate measuring machines (CMMs). The operating principles of high-accuracy interferometric measuring devices for measuring small gaps and distances between two surfaces are then described together with similar techniques for measuring small angular rotations.

The chapter concludes by discussing the application of linear and circular piezoelectric actuators for nanometric displacement. A mechatronic test bench is described which includes the implementation of a piezoelectric drive into a rotary table for angular calibration.

4.1 ACCURACY CONTROL OF THE POSITION AND DISPLACEMENT OF MACHINE PARTS

All the points of every part in a machine or mechanism are mutually interconnected. Their position and displacement can be measured and determined by selecting relevant points on the surface of the part and determining the position or distance between them in a fixed system of selected coordinates. The position of a free body in space is determined by using six degrees of freedom in the Descartes coordinate system; these parameters are named as generalized coordinates. When a body rotates about a stable axis, its position is determined by a single generalized coordinate, the angle of rotation, and during a straight-line displacement, by three coordinates of the freely chosen point on its body. The displacement of such a movement is called the straight-line displacement when this line connects the chosen point on the object's body and remains parallel to its initial position during this movement and throughout all the time of its movement.

The movement of parts of a machine or technological equipment is always restricted by mounting them in bearings and housings; their movement is also directed by slideways, and so on. Their movement can be approximated by the theoretical circular or straight-line movement, although during a more thorough analysis the real trajectory of the point on the part is assessed during its rotation or straight-line movement according to the separate axes that appear during this displacement.

Geometrical accuracy parameters of machines and instruments are usually used to assess the position of their points in the rectangular coordinate system and declinations of theoretical trajectory during their displacement. The displacement of a part of a machine is used to form the shape of the part to be machined (that is repeated many times during the industrial process) and the parts of the instrument used to check or follow the surface of the machined body. A very important machine parameter is the repeated movement into predetermined positions of some parts (positioning), for example, an instrument, in the working volume of the machine. This parameter is determined by the accuracy of the coordinate movement during multiple displacements of the part of the machine. The accuracy of the positioning is assessed in the case of straight-line or angular movement using the datum of the machine according to which the movement is performed.

The shape of a part must be machined according to the technical specifications, and for measuring equipment the shape must be controlled by determining the geometry and its declinations, that is, position, shape, length, or angle errors. The accuracy of all these parameters are implemented into the construction of the machine or the measuring equipment, based on geometrical parameters one level of higher accuracy than the production machine. Some correctional methods are applied to improve the accuracy of a machine or measuring equipment, the most popular being mathematical data processing in the control unit of the machine during machining or measuring operations. The simplest error forms to be corrected are the straight-line deviations that often occur due to environment circumstances and periodic forms of error, such as the sine form (occurring, for example, due to eccentricity of the part to be measured or being machined). Such forms are the easiest to be eliminated and can be corrected by the control unit of the machine. Apart from the main geometric accuracy parameters of the moving parts of the machines, micro- and nano-displacements have great significance for measuring machines and instruments.

Positioning accuracy (periodic position repeatability) is one of the most significant features which determines the geometric accuracy parameters for machines, metal cutting tools, and other technological equipment. A single parameter showing the coordinate displacement accuracy gives much information about the level of the machine's accuracy. Written standards for machines and metal cutting tools indicate the positioning accuracy as the deviation of the true position of part of the machine from the position that is fixed in the information measuring system (IMS) of the machine or in the programming unit that controls the displacements of parts of the machine.

This positioning error is expressed as

$\Delta p_i = p_{ti} - p_i$;
Δp_i is the deviation from the designated linear or angular coordinate position,
p_{ti} is the true value of the position of part i,
p_i is the designated (programmed) position.

The positioning of the machine parts is controlled by a control system using information from translational or rotary encoders. The IMS is a system of angular or linear encoders connected to the moving part of the machine and the nonmoving part (i.e., the base of the machine) and has a link to the control system. The measuring transducers as well as the overall IMS are checked (calibrated) for metrological consistency under conditions and regulations stated in the written standards. For positioning systems the pitch of measurement is indicated over the range of control, for example, not less than 11 measurement points in the range of linear displacement over 2 m, not less than 0°, 90°, 180°, and 270° for angular positioning. The pitch for angular displacement control is often restricted by the number of reference angle positions provided by the standard of angle measurement. For example, many angle standards, such as polygons, have only 12 flat surfaces, providing 12 angle positions for rotation positioning control. Hence the pitch for angle checking in the full circle will be only 30°. For statistical processing of measurement results at least 3 to 5 readings are taken at every position.

Standard techniques for linear measurement usually specify many more positions within the linear scale. For example, a linear scale of 1 m length may be calibrated every 10 mm or even every 1 mm. To measure the accuracy of displacement at a very small pitch of measurement is a very expensive and attention demanding process, so in many practical cases it is restricted to 10 mm calibration intervals, or more.

The fixed pitch of measurement often loses its meaning when using laser measuring devices, such as the laser interferometer for linear measurements and the laser gyro for angle measurements. The "scale" of these instruments is the wavelength of the coherent light source where the beginning and end of measurement can be chosen at every step of the interferometric pattern. The discrete value of the display unit of the laser interferometer or gyro is usually 0.1 μm or 0.1″. The accuracy of these instruments is usually given in the technical specification from the manufacturer or in certification issued by a competent metrological institution. The accuracy of linear laser interferometers reaches ~1 μm/m, and for laser gyros it reaches 1″ in a full circle. The accuracy parameters must not be confused with the technical parameters indicating the discretion of readings that sometimes are presented in technical specifications for an instrument.

4.1.1   THE ELEMENTS OF POSITION AND DISPLACEMENT

The accuracy of displacement is very important in the production of precision metal cutting tools, machines, and instruments as their accuracy is critical to their functional features. Measurement and control of parts may be performed using the simplest general purpose measuring instruments, such as sliding calipers, micrometers, gauges, indicators, and so on. The measurements and the means used for this depend on the series of parts in production and their accuracy level. The higher the accuracy of the parts the more accurate and specialized measuring instruments are required. Instruments such as Talyrond, Talyvel, Talysurf, and so on are used for roundness, straightness, and surface texture measurement. Special instruments with multisens or equipment are used for crankshafts or other shafts with complicated configuration measurement. Depending on the economic pay-off other measuring equipment may be used, that is, CMMs, instrumental microscopes, and laser scanners. CMMs can be used with different levels of automation, beginning from manually operated CMMs to fully automatic control machines, sometimes being a part of an automated production line or a flexible manufacturing system. Depending on the technical requirements, one point touch probes, multipoint measuring heads, surface tracking, or optical surface perceptible measuring heads are used. The tip of the measuring head can be as simple as a ball point or a more complicated instrument, having a measuring head consisting of several sensors or transducers, optical devices and cameras, laser scanners, and so

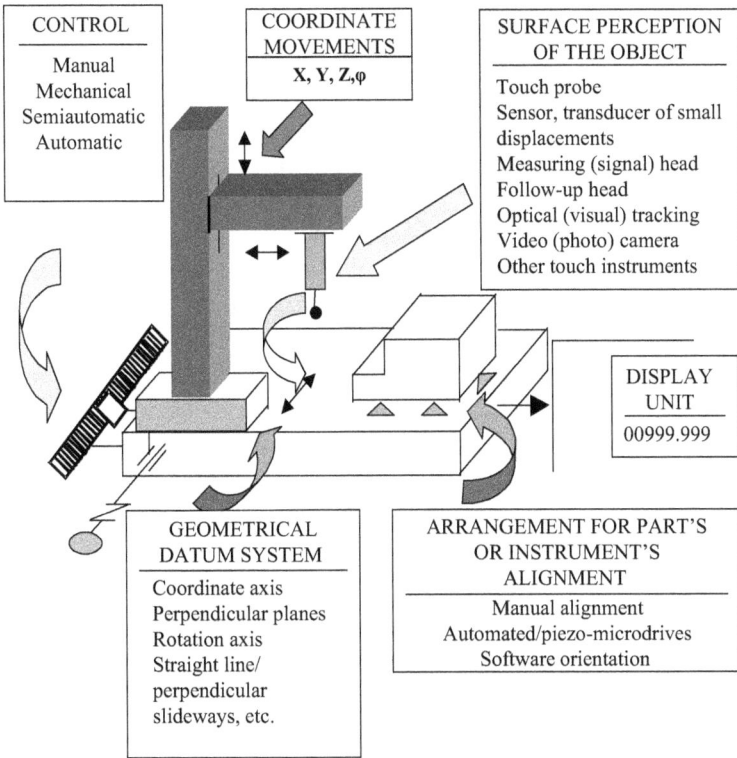

CONTROL	COORDINATE MOVEMENTS	SURFACE PERCEPTION OF THE OBJECT
Manual Mechanical Semiautomatic Automatic	X, Y, Z, φ	Touch probe Sensor, transducer of small displacements Measuring (signal) head Follow-up head Optical (visual) tracking Video (photo) camera Other touch instruments

DISPLAY
UNIT
00999.999

GEOMETRICAL DATUM SYSTEM	ARRANGEMENT FOR PART'S OR INSTRUMENT'S ALIGNMENT
Coordinate axis Perpendicular planes Rotation axis Straight line/ perpendicular slideways, etc.	Manual alignment Automated/piezo-microdrives Software orientation

Figure 4.1. Structural diagram of measuring device.

on. The measuring instrument must include the basic surfaces of high geometrical accuracy (planes, slideways) placed with high accuracy to each other, ensuring perpendicularity, parallelism to each other, so forming the datum according to which all accuracy parameters are to be measured.

A structural diagram of a measuring instrument is shown in Figure 4.1. The figure shows the generalized elements of the measuring instrument and their mutual interconnections. Basic elements of the instrument give a datum system according to which all translational or rotary displacements are performed and readings from displacement transducers are taken. The accuracy parameters of these elements must be of higher level compared to the parameters to be measured. The parts of the instrument shown in Figure 4.1 can be seen to move in three perpendicular coordinates. However, the simplest measuring device would contain only one coordinate, along which the sliding indicator is moved with the measuring tip. Such devices are used for mass production control, to check the height, width, length, or diameter of the part under control.

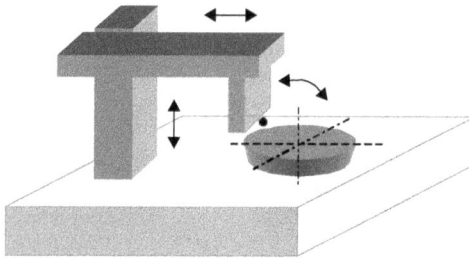

Figure 4.2. Structure of an instrument with a rotary table.

More often such devices have two or three coordinate displacements and these can also be combined with a rotary table permitting control of round parts in industry (Figure 4.2). Such a structure is the principle of many roundness measuring instruments, the rotary spindle of which has the highest level of radial error, much higher than the roundness parameters of the detail to be measured. Using two perpendicular displacements in two coordinates, the pick-up is contacted to the surface of the detail, and a measurement is taken during the table rotation.

Perpendicular slideways of the instrument permit not only transportation of the part into the measuring area, but also perform some operations of measurement, for example, to measure cylindricity of the part or straightness of the surface of cylindrical detail, perpendicularity of the end plane of the part to its axis, and so on. The displacement of the moving parts of the instrument is manual in simple instruments and mechanical for the advanced ones. Automation systems are used for displacement control, measuring the process performed, reading the results, and sending them to the memory of the instrument's control unit. Results can be presented by graph, spreadsheet, or evaluation according to tolerances given. Some data analysis is performed in modern measuring instruments, such as spectral analysis of data showing the influence of separate harmonics on the measurement results. Such an analysis can identify the factors that can influence the accuracy of the part in production. These factors can be shock and vibration acting on the technological equipment, temperature fluctuations, and sometimes humidity or pressure changes in the production area.

The IMS used in the measuring instrument is of higher accuracy-compared with the technological equipment that was used for the part's production. The simplest measuring instruments can be linear or circular scales with optical reading devices. Higher level instruments range from inductive, photoelectric, capacitance transducers to laser interferometers and scanners, holographic devices, and photo cameras with the highest digital capacities reaching tens of giga pixels. Piezoelectric transducers for sensing and actuating purposes are used extensively. Laser scanning instruments are capable of measuring surface configuration, length and angle biases in three-dimensional (3D) space. Manufacturers of these instruments claim that they can be substituted for the most powerful 3D measuring equipment, CMMs.

The surface of the part to be measured must be touched by a tactile sensor with several different kinds of end tip. A variety of tips are shown in Figure 4.3. These include ball type, cylindrical, cut-cone, prolonged cylindrical, taper cone, and many other configurations of tips.

CMMs have a changeable set of tips together with touch probes or measuring heads that are automatically changed during the measuring process depending on the measurement tasks to be performed. Figure 4.4

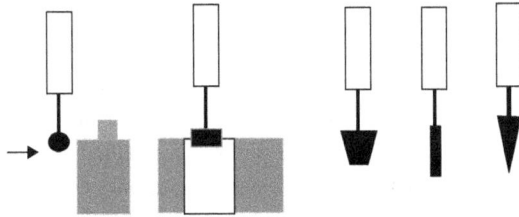

Figure 4.3. Variety of touch probes used in measuring heads.

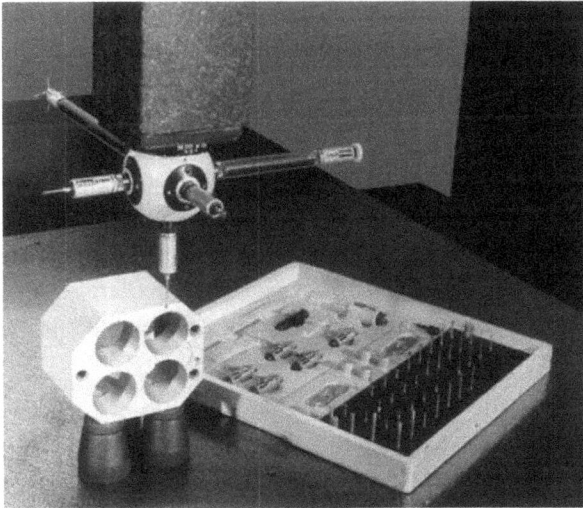

Figure 4.4. Multidirectional measuring head BE 205 (Brown & Sharpe-Precizika) of CMM with a kit of changeable tips.

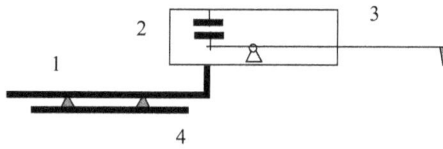

Figure 4.5. A diagram of the pick-up for measuring small (micrometers or nanometers) displacements: (1) body, (2) sensor (for example, capacitance) of small displacements, (3) touch-probe, (4) slideway.

shows a multidirectional measuring head on a CMM together with a kit of changeable tips.

A diagram of the pick-up for measuring small displacements (micrometers or nanometers) is shown in Figure 4.5.

In the body of the measuring head there is a primary sensor (inductive or capacitive) with its sensing part connected by the touch-probe to the surface of the part to be measured. The sensing device can move on the slideway by means of which the tip of the probe is connected with the measuring surface. Such a general structure is used for roundness, surface roughness, and other measuring instruments of geometrical parameters. By touching the measuring surface with the tip of the device, it becomes sensitive to changes of the tip position from the initial readings. The tip movement is transferred to the primary sensor; it is amplified, converted into digital form, and transferred to the display. Sometimes for accuracy monitoring purposes only the signals (+) or (−), "fit," "nonfit," red, or green are used. The signal output is transferred into the control unit of the production line.

Many modern measuring devices have optical or photoelectric noncontact sensors instead of contact probes. Noncontact probes also follow the surface of the part to be measured and their position according to other surfaces or the datum of the measuring device. In some cases these functions can be performed by the operator using the optical system of the instrument. Of course, optical digital cameras with charge coupled device (CCD) matrixes and sensors are convenient as possibilities for automation and data processing using microcomputers or processors.

4.2 MEASUREMENT OF NANOMETRIC PARAMETERS

High-accuracy interferometric measuring devices were developed for the measurement of the contact point of touch probes of machines and

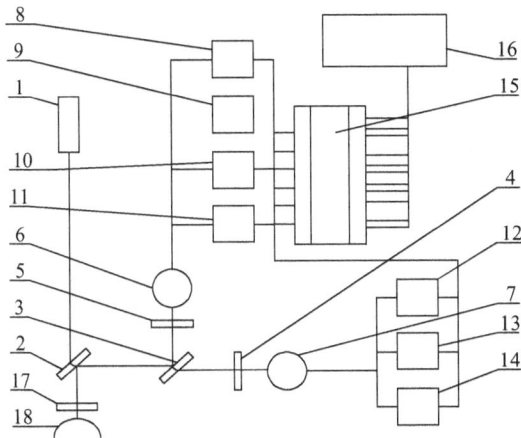

Figure 4.6. Interferometer device for the measurement of the distance between two surfaces.

measuring equipment. A diagram showing the principle of measurement of small gaps and distances between two surfaces is shown in Figure 4.6.

A two-frequency light source (1) is used in the device. The light beam goes through the split plates (2) and (3) where it is split into two beams. Light filters (4) and (5) transmit light of a single frequency. The light beam proceeds to the photodetectors (6) and (7), where it is converted into digital form by the comparators (8–14) and the coding loop (15). The coding coincides with every discrete combination of the light signal coming between the surfaces of the objects (17) and (18). The result of the measurement is registered in the display (16). The light beam interferes after reflection from the transparent surface (17) and opaque surface (18). The output of the photodetectors can be described by the trigonometric functions which characterize the distance between the two surfaces to be measured.

$$
\begin{cases}
U^{(1)} = 2(1 - \cos\dfrac{4\pi}{\lambda_1} h) U_0^{(1)} \\[2mm]
U^{(2)} = 2(1 - \cos\dfrac{4\pi}{\lambda_2} h) U_0^{(2)}
\end{cases}
\tag{4.1}
$$

where $U^{(1)}$, $U^{(2)}$—signals at the output of the photodetectors,
$U_0^{(1)}$, $U_0^{(2)}$—signals of stable value, dependent on the optical part of the device,
λ_1, λ_2—the wavelength of the light transmitted through the photofilters;
h—distance to be measured.

The output of the signals can be expressed through the wavelengths of the light λ_1 and λ_2, as well as by their parts $\lambda_1/4$, $\lambda_2/4$; $3\lambda_1/4$, $3\lambda_2/4$, and so on. The difference between the signals in the first half of their periods will be $(\lambda_2 - \lambda_1)/4$, in the second $2(\lambda_2 - \lambda_1)/4$, the third $3(\lambda_2 - \lambda_1)/4$. After n half periods the difference will be $n(\lambda_2 - \lambda_1)/4$. It means that a single determination of the distance is possible, using the lengths mentioned above when the halves of the periods coincide, or

$$n(\lambda_2 - \lambda_1)/4 = \lambda_2/4;$$
$$\text{or } n = \lambda_1/(\lambda_2 - \lambda_1).$$

The range L of the distance h determined will be

$$L = \frac{\lambda_2}{4} \cdot \frac{\lambda_2}{\lambda_2 - \lambda_1}.$$

The signals of analog form are converted by the comparators into digital form. The number of comparators must be selected such that the pitch of the range of measurement would be n-times less than a half period of the signal

$$(\lambda_2 - \lambda_1)/4, \text{ where } n = 1,2,....$$

A half period of the analog signal $U^{(1)}$ is divided into the number of steps

$$K_1 = \frac{\lambda_1}{\lambda_2 - \lambda_1}, \text{ and the second half period into } K_2 = \frac{\lambda_2}{\lambda_2 - \lambda_1}, \text{ parts.}$$

The logic signal, formed starting with the last half-period division step, is always equal to zero. Thus, comparator number m is less than number K by one, that is

$$m_1 = K_1 - 1 = \frac{2\lambda_1 - \lambda_2}{\lambda_2 - \lambda_1}. \quad m_2 = K_2 - 1 = \frac{\lambda_1}{\lambda_2 - \lambda_1}.$$

The cumulative comparator number will be

$$m = m_1 + m_2 = \frac{3\lambda_1 - \lambda_2}{\lambda_2 - \lambda_1}.$$

The comparator operating level is calculated in a step-by-step manner from the above formulae, inserting the values of h. The sequential number of a z-step shows the magnitude of a gap and is

$$h = z\frac{\lambda_2 - \lambda_1}{4}.$$

The signal on the output of the coder is the output according to the real complex of signals from the comparators. Its sequence number coincides with the sequence number of the step. In such a way the information about the step number is presented.

Another method for the measurement of small lengths, gaps, and distances between two surfaces operates as follows. This method enables the range of measurement to be expanded to several micrometers. The resolution of measurement remains very high, at least 0.05–0.005 μm. The principle of operation of the device is shown in Figure 4.7.

The light beam from laser (1) is focused into the surface of the probe's point (19) through the transparent plate (18) by the lens (17). Light waves reflecting from the probe's point and the plate's surfaces interfere and so interference waves are formed. The interference waves go through the semitransparent mirrors (2) and (3), and to photodetectors (6) and (7). The light filters (4,5) convert the light pattern into a monochromatic one. Signals coming from the photodetectors are as the ones mentioned above and described by those equations as well. The discretization is also performed in the same way.

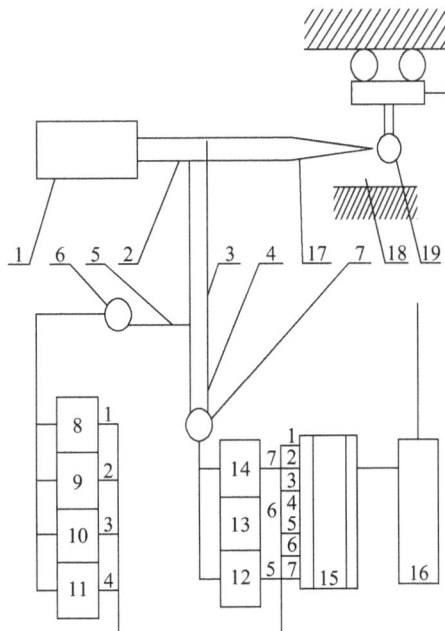

Figure 4.7. Diagram of measurement of the distance between the probe's point and the surface to be touched.

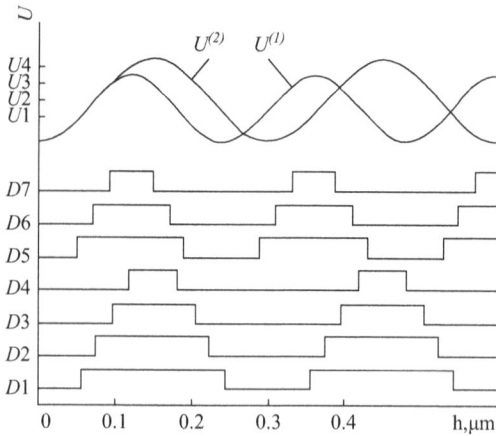

Figure 4.8. Diagram of the formation of the discrete code.

Figure 4.8 shows the initial position of the former analog signals, and the lower diagrams demonstrate the rate of their changing in the form of a discrete digital code. Analysis of the signals made it possible to describe the values of the voltages in the photodetectors, $U^{(1)}$, $U^{(2)}$, in terms of a discrete signal set in the comparators, $D1$–$D7$, according to the distance being measured. The signals of analog form are converted into digital signals $D1$–$D7$ in the comparators (8 to 14). Signal Di is equal to one, if the voltage in photodetectors (6) and (7) is higher than the preset level of the voltage. Signal Di will be zero, if the voltage in photodetectors (6) and (7) is less than the preset level.

The measurement is performed by the same channel in sequence by modulating of the light beams with the wavelengths λ_1 and λ_2. The signals in photodetectors are

$$\begin{cases} U^{(1)} = U_{g1} + 4U_{01} \cos^2(2\pi h / \lambda_1), \\ U^{(2)} = U_{g2} + 4U_{02} \cos^2(2\pi h / \lambda_2), \end{cases}$$

where U_{g1}, U_{01}; U_{g2}, U_{02} are constants of the first and the second signals, respectively, depending on the design of the device; h is the distance to be measured.

The result of measurement is cumulated in the indicator unit (16) according to the signal delivered from the encoder (15). The sequential change of the processed signals enhances the singularity of the value of measurement, as the signals cannot be coincided, interchanged, or moved during the processing.

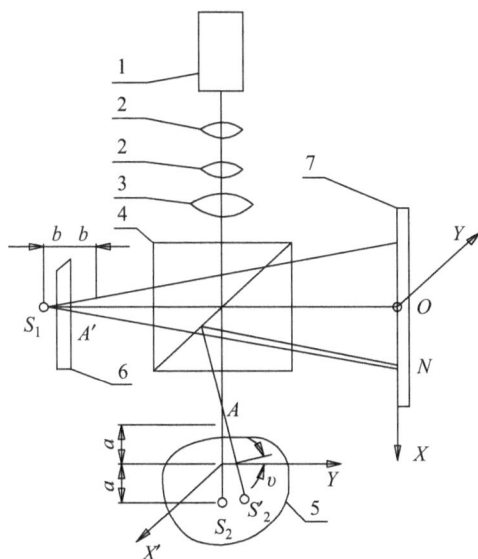

Figure 4.9. Method of measurement of small angular rotation.

For the purpose of presenting the measurement information in analog form, the device can be rearranged in a suitable manner. The technical solution makes it possible not to lose the measurement information both in the range of measurement, and in the different parts of it as well. There the best advantages of discrete and analog signal processing are used. The larger lengths are measured by using the discrete transformation, and the small distances left are added to the common analog signal.

A method for measurement of small angular rotation is presented in Figure 4.9. The essence of the method is the formation of spherical laser light waves from the reference surface and from the surface to be measured. The light beam from the laser (1) through the collimator (2) and lens (3) is focused into point A. The light split plate (4) splits the beam into two parts, one part reflecting from the reference mirror (6), and the other part from the mirror (5), fixed with the object, small rotation angle of which is to be measured. Coherent waves of light reflecting from the mirrors have spherical fronts of different radii. By interfering on the screen (7), they form ring-shaped fringes. The angle of the rotation of the object can then be determined from the shape and dimensions of the fringes. The method can be used for the measurement of angular rotation along the three rectangular coordinates and angular shifts in two rectangular planes, that is, pitch and roll.

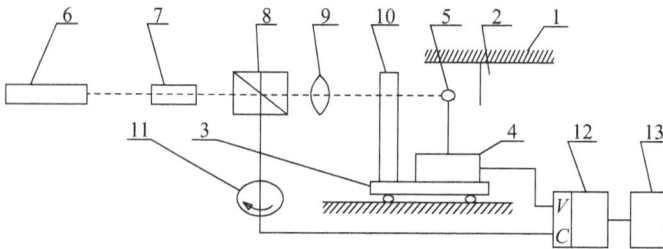

Figure 4.10. Simplified diagram of interferometer device for IMS accuracy measurement.

Control of the accuracy of an IMS by interferometric means is described below. The main purpose of the measurement is to determine the value of $dXi = Xi - Xo$, where Xi is the coordinate position value fixed by the IMS and Xo is the real coordinate position at the moment of touching the part surface by the probe or the machine instrument.

The simplified diagram of the device for IMS accuracy measurement is shown in Figure 4.10. The measurement is performed by using the light beam from interferometer (6), transmitting through the collimator prism (7), split plate (8), lens (9), transparent plate (10) to the point of the probe (5). In this case, when the point of the probe does not touch the surface of the support (2), the distance between the probe point and the support has no changes and the signal of interferometer keeps the counter (12) closed and input of the signals from photodetector (11) is prohibited. When the point of the probe (5) touches the support (2), the distance between the plate (10) and the probe point changes and the signals are permitted to enter the counter. Due to the systematic error of the probe and the IMS, the counter input is closed after some delay by the signal from the IMS, and the indicator unit (13) keeps the result of the IMS error.

The signal from the photodetector is in fact of random value, but remains constant until the moment of touching the surface.

$$U = U_0(1 - \cos\frac{4\pi}{\lambda}h),$$

with λ—the length of light wave,
         h—initial value of the distance between surfaces,
         U_0—constant, dependent on the design of the device.

The point of the probe touches the support at the point X_c, the distance change being

$h + X - Xc$, and the signal output from the detector is

$$U = U_0 \left[1 - \cos\frac{4\pi}{\lambda}(h + X + X_c) \right].$$

The interferometer device constructed for this purpose was investigated and the results of measurement did not exceed the value of 0.12 ± 0.05 μm at $P = 0.95$.

More valid results can be achieved by using the device shown in the diagram in Figure 4.11. Here two measuring signals are used, optically linked with the moving part (carriage) of the machine and with the probe.

The measurement system can be accomplished using one laser and optical split plate or two different laser units for each channel of measurement. One channel of measurement goes from unit (6), reflecting from carriage (2), another one from unit (7), reflecting from the point of the probe (4). Both light beams interfere with the reference beam. The carriage moves on the bed (1) to the part to be measured (5). As was described

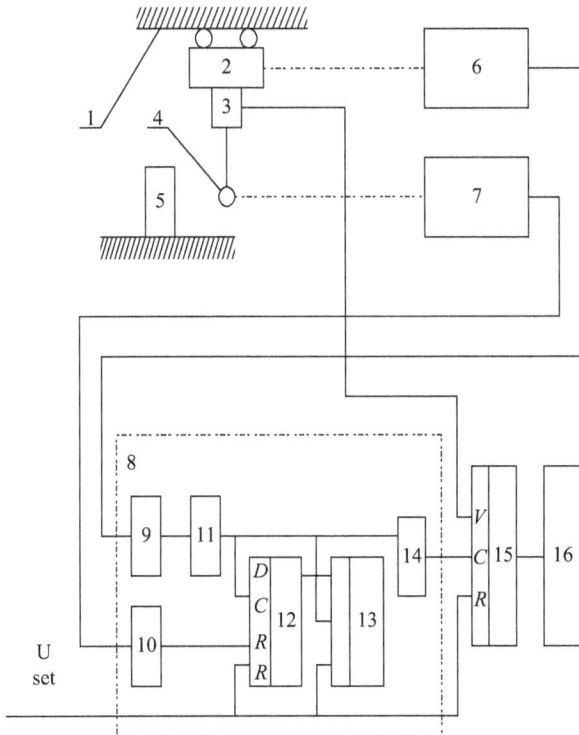

Figure 4.11. IMS measurement using two interferometer signals.

above, there is no signal from the probe, while there is no contact between the point and the surface of the part. There is no output signal from the unit (8), which is reset to "0" prior to measurement, as well as the result in the counter (15). When the point of the probe touches the surface, it is stopped, and the signal output from the channel (7) opens the counter (15) for cumulating the measurement information from the other channel. This counting will be interrupted by the signal output from the transducer (3). The cumulated result gives the true value of the IMS error. The usage of the same IMS for functions of machine performance and for auxiliary functions (changing the probes or instruments in the FMS environment) is available.

4.3   LINEAR AND CIRCULAR PIEZOELECTRIC ACTUATORS FOR THE NANOMETRIC DISPLACEMENT

The application of piezoelectric actuators in a comparator for angular measurements is shown in Figure 4.12. Here a plan view is added showing the use of piezoelectric actuators for rotary table positioning at the pitch of measurement predetermined by the measurement program. The comparator consists of the housing (1); rotary encoder (2) as a reference angular measure; (3) elastic clutch; (4) rotary table; (5,6) photoelectric microscopes; (8) circular scale to be measured; (7) strokes of the scale; (9) generator of high frequency voltage oscillations; (10,11) piezoelectric plates; (12) plates of hard plastic; (13) elastic suspensions.

The rotary encoder (2) is connected with the rotary table via the coupling (3) and the table (4) is mounted concentrically. The table (4) is equipped with an adjustment device for adjustment of the circular scale (8) in the radial direction and for leveling its flat surface in the horizontal direction perpendicular to the axis of rotation. Photoelectric microscopes are pointed above the strokes (7) of the scale. The microscopes (5 and 6) are placed on diametrically opposite sides in order to eliminate eccentricity of the scale during measurement.

One of the main requirements for such type of equipment is the accuracy of rotation of the table on the axis. This task was efficiently solved by using piezoelectric plates (10 and 11) fixed by elastic suspensions (13) to the rotary table. The symmetrical position of the plates and the elliptic movement of the contact point to the table enable rotation of the table with the highest accuracy, not distorted by the driving mechanism. Trials confirmed the high accuracy of the table using piezoelectric actuators. The run-out of the axis under rotation was within the values 0.1 to 0.05 µm,

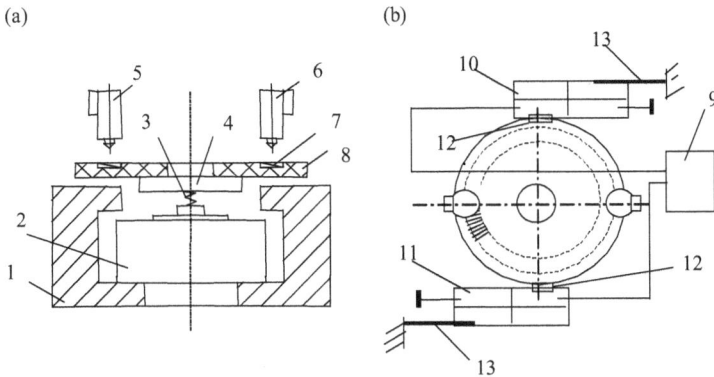

Figure 4.12. Diagram of angular comparator (a) and its plan view (b) with piezoelectric actuators.

and its axial value didn't exceed 0.05 μm. The piezoelectric plates are excited by high-frequency voltage oscillations from the generator (9). The mechatronic control of the actuator enables the rotational velocity of the table to be changed within wide limits. Very slow adjustment movements are also available.

The main technical specifications achieved are as follows:

- velocity of rotation, ω, rev/min   0.1–10;
- axis run-out, μm   0.1–0.05;
- reversibility   yes;
- automatic control   by PC or microprocessor.

The mechatronic comparator for calibration of circular scales and rotary encoders is controlled preferably by PC as the data received during the calibration is very extensive. The feedback for the comparator's control can be received from the reference encoder's signals for data transfer from the photoelectric microscopes (5 and 6). The manufacturing process of raster scales and transducers is such that some systematic errors are common. The cause of such errors might be a technical fault in the production of the primary (original) scale, faults during the calibration process, such as shock and vibrations, temperature changes, and so on. Errors often occur at the beginning and at the end of the raster scale. In such cases, special methods for accuracy correction of the raster scale and the transducer are available.

The accuracy of positioning of the reference measure and the measuring object can be improved by selecting a mechatronic impact on the information-measuring system for the elimination of systematic errors.

Piezoplates are applied for micrometric displacement to correct the systematic errors of the scales or transducers of the information-measuring system of the machine. Precision displacement is accomplished by control of the potential energy of the piezoplate or other element of the piezo-actuator. From a technical point of view this can be done by changing a voltage supplied to the electrode electromechanically coupled with the part of the transducer or changing the width of the electrode coating of the piezoplate.

A generalized mechatronic measuring system is shown in Figure 4.13 as used for angle calibration of measuring equipment. It is useful for circular raster and coded scales measurement in conjunction with the scale's strokes reading instruments, such as photoelectric microscopes or autocollimators for the determination of the angular position of the object. Information of rotation at the predetermined step of measurement can be controlled in an automated way when the system is controlled by the output signal of the rotary encoder. The rotary encoder also gives the information of the position at which a systematic error is determined for further correction. This can be done by changing a voltage supply to the piezoelement selected as an actuator for adjusting the position of the microscope or other reading instrument of the measuring equipment.

The main constituents of the measuring system are the reference angle measure (rotary encoder of high accuracy), the high precision air bearing (uncertainty of run-out is ~0.05 μm), piezoelectric drive for movement of the measuring object at a predetermined step of measurement, rotary

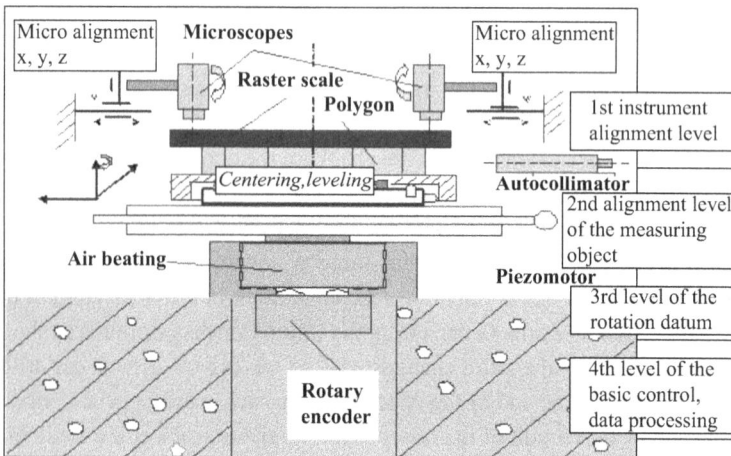

Figure 4.13. General diagram of measuring equipment with mechatronic structure.

adjustable table (with the possibility to be adjusted in at least three degrees of freedom), the reading devices to measure the angular position of the object (photoelectric microscopes or autocollimators with CCD matrixes), fixtures for adjustment in the radial, tangential and angular directions, signal outputs from all active elements into the PC, hardware and software for system adjustment, control of operation, readings transfer, result processing and presentation.

The objects to be measured are the circular raster scales, rotary encoders, and coded scales. The task of the measuring system is to determine the bias of the circular scale angle measuring standard and to use the results for error correction and accuracy improvement of metal cutting machines, CMMs, robots, and so on. This is achieved with the application of active materials such as smart piezoactuators implemented into several positions of the measuring equipment. The mechatronic measuring system is analyzed as a complex integrated system and some of its elements can be used as separate units.

The table for centering and leveling of the object under the reading microscopes must be able to perform an alignment displacement in the range 20 to 50 μm with a resolution of ~0.1 μm. Only piezoelectric actuators are suitable for such a task.

Calibration equipment for angle measurement includes components from various branches of technology, including optics, electronic devices, drive units (vibromotors or piezoelectric or step motors), digital information links and data processing, including computer control of all the processes. A typical mechatronic test bench would include photoelectric microscopes and autocollimator, an optical angle standard (multiangular prism/polygon), a rotary table, driven by a step motor or piezodriveand controlled by a photoelectric rotary encoder and PC. The PC serves as the main control unit, operating not only for the control of the measuring process, but also for data selection and processing operations, including mathematic statistical evaluation and presentation of results in the form of digital protocol or diagrams of error distribution.

A real implementation of a piezoelectric drive into a rotary measuring table is shown in Figure 4.14(a). A construction with a more usual motor, for example a step motor, is shown in Figure 4.14(b). It is evident that the piezodrive has many advantages compared with the drives in general use [27]. In fact, the piezodrive consists of a single piezoelectric plate (Development of R. Bansevičius, V. Jurenas, Kaunas University of Technology, Lithuania) connected to the voltage supply from a generator, where the value of the electricity supply and the frequency of voltage can be controlled. Mains supply in the first and second mode of oscillation gives an oval

(a)   (b)

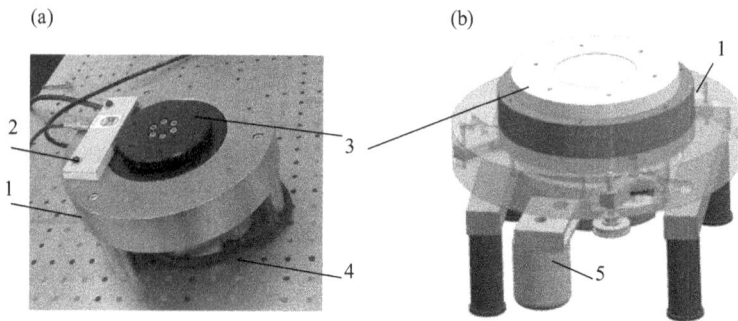

Figure 4.14. Experimental rotary table with mechatronic parts: (a) real implementation with piezodrive; (b) version of modeling applying a step motor.
1—base of the table; 2—piezoelectric drive; 3—air bearing of high precision
(uncertainty of run-out ~0.05 µm); 4—angle standard—photoelectric rotary
encoder of high accuracy; 5—step motor.

form of trajectory of the contact point of the piezoplate, thus rotating the
disc of the table.

The main dimensions of the rotary table are determined by the dimensions of the rotary encoder and the air bearing; the piezodrive used for
rotation occupies almost no height in the total construction of the arrangement. This is an important advantage of the measuring device, complying
with the requirements of the Abbé rule to keep the measuring line or plane
as close to the reference measure as possible and so avoiding the influence
of the Abbé offset.

CHAPTER 5

TESTING AND CALIBRATION OF COORDINATE MEASURING MACHINES

This chapter covers the testing and calibration of coordinate measuring machines (CMMs). The operating principles and applications of CMMs are first discussed, followed by the principles of measuring heads (MHs) and touch trigger probes. The performance verification of MHs and touch trigger probes are described including the use of new artifacts.

Methods for the full geometric accuracy calibration of CMMs are covered, including laser calibration against a reference or master machine. Performance verification using both artifacts and lasers is also covered in detail. Finally the design of error correction mechanisms using piezomechanical correction techniques is proposed.

5.1 INTRODUCTION TO PRINCIPLES OF OPERATION AND APPLICATION OF CMMs

The CMM is used for manual or automatic measurement of the geometric parameters of parts in production. These parameters include the following:

- the overall length, width, and height;
- the distance between points on the part's surface, inner and outer circumferences, and coordinates of the axis of cylindrical or spherical surfaces;
- parameters of coaxially, cylindricity; perpendicularity of the lines and planes on the surface of the parts, and so on.

It is a sophisticated measuring instrument, usually mounted on a granite table with other parts and the probe suspended on the ram for generating a signal that measures parts in three-dimensional (3D) space. Usually a CMM has three coordinate axes, but sometimes it is supplemented by a rotary table, so expanding this to four coordinate axes. Some axes for the MH's tip orientation can be implemented into the construction of the MH.

A widely used construction of CMM is the bridge type where the bridge is moving on a table supported by air bearings. The carriage moves along the bridge and inside, in a vertical direction moves the ram, with the MH (touch probe) suspended on it.

The cantilever-type CMM is supplied with the single moveable vertical support that suspends a horizontal arm with the probe. Such construction makes it easier to access the staging table and the part on it. The gantry type CMM is similar to the bridge type, only containing a much larger part, the gantry. It is used for measurements of large components in industry, that is, in the aviation industry, car industry, and so on. A portable CMM is used for measurements in a manufacturing plant where it is taken to the part to be measured. Most modern CMMs are controlled by a PC using special software for control of the measuring process and processing of results.

Before taking the measurement the process of relating the part coordinate system to the machine's coordinate system must be performed. The alignment allows the part to be positioned anywhere on the staging table. A point, or geometric features of reference from which measurements are taken, are treated as datum, for example a hole, a line, or any other geometrical shape.

The typical "bridge" CMM is composed of three axes, X, Y, and Z (Figure 5.1). These axes are orthogonal to each other in a 3D coordinate system. Additional coordinates can be added, for example, axis of the rotary table. Straightness, flatness, perpendicularity of movement of the parts of the CMM constitute the datum, that is, the theoretically exact geometric reference parameters that ensure the basic features against which the geometrical parameters of the part are measured on the machine. Each coordinate axis has a displacement transducer system (information measuring system [IMS]) that indicates the location in that axis. The machine will read the input from the transducer after the command of the touch probe during the contact of its tip with the surface of the part to be measured, as directed by the operator or the control program. The machine then uses the X, Y, Z coordinates to determine the position of the point in the coordinate axes. Typical discretion of a CMM is measured in microns (micrometers), 1/1,000,000 of a meter. The accuracy of measurement depends on the measuring volume and differs in the range of several to several tenths of microns.

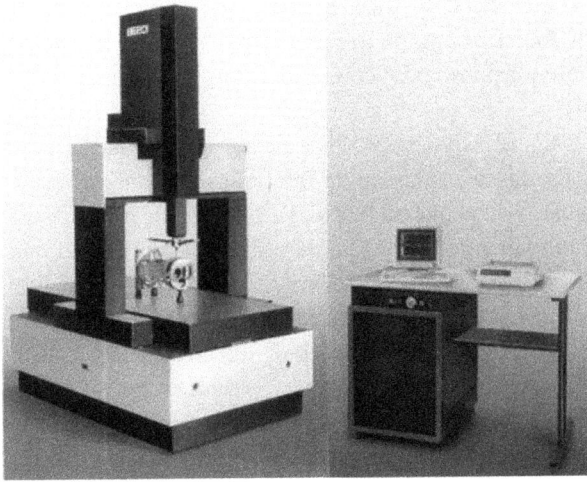

Figure 5.1. Coordinate measuring machine mod. BE 201 (Brown & Sharpe Precizika).

A CMM is also a device used in manufacturing and assembly processes to test a part or assembly against the design drawing or technical specifications. By precisely recording the X, Y, and Z coordinates of the target, points are generated which can then be analyzed via algorithms for the construction of features. These points are collected by using a probe that is positioned manually by an operator or automatically via computer control. Computer controlled CMMs can be programmed to repeatedly measure identical parts; thus a CMM is a specialized form of industrial robot, sometimes called as a measuring robot. The part for measurement is placed on the table of the CMM and aligned along the coordinate axis of the machine. The coordinate system of the part is created by the software and transferred to the coordinate system of the machine. The surface of the part probing is performed by discrete points or during continuous tracking of the surface. Discrete measurement is when the touch trigger probe contacts the surface and generates a signal at this point, the coordinates of which have been entered into the memory of the PC. Continuous tracking (scanning) is a form of measurement when the MH keeps continuous contact with the surface of the part and during the fixed periods the position of the surface points is entered into the PC memory. After the selection of information about the coordinates of the points that were probed, the results of measurement are developed, determining the overall dimensions of the part, form and deviation of the surface, position of the holes, hole diameters, distances between the axis of the holes, determination of the tolerances of the deviations measured or providing the points and distances that are to be corrected during the manufacturing process, and so on.

The main parts of the CMM are the housing, the bridge moving along the base, the carriage moving on the bridge and the ram which usually moves in the vertical direction (z axis). Usually, the datum of the machine is made from a granite base. Measurements are defined by a probe attached to the ram of the machine and the probes may be mechanical, optical, laser, or white light. The principle of operation of the CMM may be either contact or noncontact. Mechanical or automatic CMMs have electromechanical drives and use an encoder system as the IMS that gives the coordinate position at the moment when the probe of the CMM touches the surface of the part. Linear encoders, rotary encoders, or laser interferometers (LIs) are used for the IMS of the machine.

The most advanced methods can be used for the assessment of the accuracy parameters of the CMM due to its usage for highest accuracy quality control in research and industrial production. Different optical methods applied for 3D measurements are investigated, including the projected fringes method, electronic speckle interferometry, structured lighting reflection techniques, white light interferometry, and laser scanning methods. The instruments for fast 3D surface points detection and numerical presentation are continuously under development and the advantages of fiber optics, fiber sensors, and measurement systems based on them are utilized. The use of optical methods for displacement and position measurements gives great advantages, such as minimal dimensions, wide range of accessibility, integration of the sensor and information transfer functions in one unit, and so on.

It should be noted that these modern methods are not yet used widely enough in CMM calibration and the accuracy verification processes of Computer Numerical Control (CNC) machines. Special reference parts (artifacts) are commonly used for CMM calibration purposes with their entire dimensions assessed beforehand with high accuracy. Many investigations are made to determine the measuring coordinate system uncertainty using datum planes, cylindrical and spherical bodies, sets of disks placed, and calibrated in the measuring volume of the machine. A variety of configurations of such artifact parts are used. They include rectangular plane surfaces, internal cavities or exterior spherical bodies, cylindrical and conical surfaces, and so on. The disadvantages of such methods of accuracy control are in the restrictions of the measurement volume of the machine to be assessed.

Most of such reference parts have relatively small dimensions as it is inconvenient to handle a reference part of large dimensions on the table of the machine. Therefore the calibration does not always cover the complete volume of the machine and a large part of the working volume of the machine is unavailable for calibration. Consequently, there is a need to investigate, develop, and introduce into practice, some new methods which can be easily and conveniently used for the calibration of accuracy parameters of machines.

All geometrical accuracy parameters are multiplied, splitting them into the coordinate axis of the machine, so making the checking more complicated. For the accuracy verification of CMMs it is important to choose the etalon, or reference measure, for each of the parameters, for example, a reference length measuring system as the standard measure to calibrate all three coordinate axes of the CMM. The precision length measuring system is chosen from a wide variety of reference measures used for this purpose. From the variety of length standards, an LI is used most often as a reference measure. Together with high accuracy of the reference measure, there are other requirements to be followed during the accuracy verification of CNC machines, one of them being the importance to avoid the influence of Abbé offset, cosine error and tilting errors, occurring during the measurements. It is important to determine errors which act in the different directions of coordinate axes in the measuring volume. For many machines the measurements are performed not only into the directions of the coordinate axes, but also in the diagonal direction of the measuring or working volume of the machine. Diagonal method is efficient because of the involvement of three dimensions in every step of the measurement performed. There are some methods of calibration where the redundancy technique is employed. The calibration in the area or volume is performed by placing the reference measure at several different positions. An optical scale can be used for such purpose. The systematic errors of the machine are identified and after mathematical analysis the errors are corrected by some technical means.

The problem remains to develop calibration methods that gather a large quantity of accuracy data in a short period of calibration, using simple reference equipment of low cost. The next step would then be the efficient use of data selected for the accuracy improvement of the machines. Some methods used to achieve these objectives are discussed and applied in monitoring operations.

Traceability and calibration of the accuracy parameters of the parts of multicoordinate machines, robots, and CMMs is quite a complicated task to accomplish and assess. The large amount of information in 3D measurements and the lack of universal spatial reference measures are the main technical and metrological tasks to overcome. Some technical methods for the measurement of accuracy elements are presented here and some nontraditional methods for complex accuracy assessment are proposed.

The accuracy of geometric parameters of precision machines, metal cutting tools and CMMs, is quite a complicated task to assess because of the variety of accuracy parameters to be checked and the high accuracy that must be assured. It is a time-consuming process that requires very experienced and well-qualified staff to perform the calibration and several reference measures to achieve the traceability of the calibration. The vast amount of information in a 3D measurement volume and the number of reference measurements needed to cover all the range of measurement space are the main technical and metrological tasks to overcome. CNC machines have some additional features, which add to the complexity of overall calibration, such as the accuracy parameters of the rotary table, the position of the cutting instrument and the instrument changing devices. For CMMs the additional features are the accuracy parameters of the MH (touch trigger probe). The methods of measurement of the accuracy parameters of CMMs are summarized in Table 5.1 later in this chapter.

The position of the moving parts of CNC metal cutting tools and other machines is determined by geometric accuracy parameters in respect of six degrees of freedom (DOF). These accuracy parameters consist of a separate group of technical parameters of the machines that need separate and sometimes rather special means of measurement for testing or monitoring. The six DOF are often referred to as generalized coordinates. These parameters are described by technical specifications on the machine and special written standards for separate parameters or for total accuracy verification. A set of instruments is used with different classes of accuracy (higher, of course, than the accuracy of the geometrical parameter to be checked), special ranges of measurement, sensitivity, resolution capability, and so on. Modern machines have many coordinate movements, much more than six of them, so the number of DOF will also be higher. Some parameters are too complicated for measurement and sometimes there is no means of measuring one single parameter without the influence of the others. An example is the measurement of straight-line movement of part of the machine by the use of a reference measure of the straight line in the form of a flat surface of the reference measure made from steel or granite. A contact inductive gauge usually is used for the measurement of this parameter, and during measurement the other parameters, such as the pitch of movement of the moving part also influences the measurement results. In the case when more exact accuracy analysis is needed, or when there is need to investigate separate accuracy parameters, then additional measurements are required using different methods. The measurement mentioned above must be repeated using an autocollimator and reflecting mirror, which is not sensitive to displacements perpendicular to the movement direction.

These and other technical specifications assure very high accuracy of geometry and high smoothness of the surface of the machined workpiece. For example, run-out of a cylindrical surface can be achieved to an accuracy of 0.2–0.3 µm. Also the control unit of the machine feeding a grinding wheel transfers an accuracy of the linear displacement transducer (LDT) to the required value that can be controlled in two coordinate directions. Widely used LDTs for CNC machines are photoelectric raster scales incorporating a reading head which, after an interpolation, assures an in feed value to within 0.001 mm (or even 0.0001 mm). Geometrical accuracy parameters also include the high geometric accuracy of slideways, their mutual parallelism or perpendicularity, coaxiality of front and back grinding heads, and so on. All these parameters are to be checked during machine manufacture and then during operation of the machine in industry. Checks are carried out according to the written standards on the relevant metal cutting tools and also according to technical specifications on the machine. Many accuracy parameters are similar for various types of metal cutting tools, such as milling, grinding, coordinate drilling and grinding, and for various types of measuring equipment, such as instrumental three-coordinate microscopes, roundness measuring instruments, CMMs, and so on.

Machine performance tests are held at certain periods or under special requirements, for example, after evident failures, accidents, functional failures, or loss of accuracy. A predetermined number of procedures are foreseen for this purpose. The parameters to be tested can be illustrated on CMMs or grinding machines, as they feature the highest accuracy parameters and a wide variety of DOF are used in their design.

EN ISO standards [14] specify the following CMM tests:

- acceptance tests, during which the performance parameters of the machine and measuring system comply with the technical specifications declared by the supplier;
- reverification tests (periodic inspection), during which the user has an opportunity to periodically check a machine's performance parameters and the measuring system;
- interim tests during which the user can check the performance of the CMM and its probing system between regular reverification tests;
- calibration, performed as a full parametric calibration.

Accuracy tests are included in practically all kinds of tests as accuracy parameters have the most importance influence on the machine's performance ability. They are undertaken using length end blocks and gauges, metal linear scales with microscopes, and LIs. The tests are performed in various positions in the working volume of the machine, using the techniques discussed in Chapter 2.

Before the beginning of tests the CMM must be put into operation according to the manufacturer's recommendations, the MH is identified and its compatibility checked. The working and environmental conditions must be set to comply with the technical specifications on the machine. All performance verification is tiresome and is a time-consuming process, so it is necessary to search for time savings, efficiency, and reduction of costs.

Minimal requirements can be fulfilled using some machine monitoring tests. The use of artifacts for machine verification can be considered as one of the possibilities for machine monitoring processes. The main task of these procedures is to keep equipment at the required level of accuracy and at its proper performance. The tests are performed under a mutual agreement between the manufacturer and user of the machine. Basic machine tests are used to check how the length measurements are performed on the machine. The error E is checked by measurement of length gauges.

Length (positioning) measurement. This is one of the accuracy tests within CMM performance verification. It is performed by measuring the accuracy of linear displacement in three coordinate axes using the following material standards of length, gauge blocks and step gauges, reference linear scale or LI. The measurement is used to check the traceability of CMM length measurements to the international standards of length. Different length measures are recommended for this test. It is recommended that:

- the longest length of the material standard is at least 66 percent of the longest diagonal of the working volume of the machine;
- the shortest length of the material standard is less than 30 mm.

If a user's material standard is used, the error must be not greater than 20 percent of value E. If it is greater, then E must be redefined as the sum of E and this uncertainty. In the case of the use of a manufacturer's material standard, then no additional uncertainty shall be added to the length measurement value E.

Measurements are performed in every direction of the working volume of the CMM at the user's discretion, making bidirectional measurements either externally or internally. Five test lengths are chosen, each measured three times. The error of length measurement is calculated as the absolute value of the difference between the indicated value of the CMM and the true value of the relevant test length. Gauge blocks (Figure 2.6) are widely used as material length standards for this purpose. Systematic errors can be corrected if suitable software or instrumentation is available in the CMM construction. Supplementary measurements are usually made usually during the alignment process.

In mechanical engineering (metal cutting tools, CNC machines, and instruments) measurement intervals should not be longer than 25 mm while measuring a displacement of 250 mm or less. For longer travels, up to 1000 mm, the interval should be no shorter than 25 mm or no longer than one tenth of the length of travel. For travels longer than 1000 mm, the measurement interval should be no longer than 100 mm. The measurement interval of the length so that no fewer than 20 positions could be measured is indicated in CMM documentation.

Repeatability measurements should involve an assessment of all aspects of the machine, including the mode of operation, work of the operator and software performance. The test conditions should be as close to the working conditions as possible. Hence there are requirements for different tests to be undertaken using different modes of operation. A modified test procedure must be applied for a machine with a large working volume and in special cases, which do not meet the requirements of the standard, the supplier and user should agree on the selection of tests to be completed.

Measurement can be performed using end-length measures and an MH. At the first point of contact by the end-length measure the readings of the machine must be set to zero. During each travel three measurements must be performed and the average arithmetic result is calculated. The process is repeated in all three coordinate axes. To minimize the influence of the environmental conditions on measurement, all measurements should be performed in as short a time period as possible.

To determine the measurement uncertainty of 1D, 2D, and 3D measurements, one, two, and 3D measurements are employed. They differ in the number of directions in which displacement is performed, and the corresponding coordinate measurement systems are at work during measurements. The uncertainty of 1D length measurement is checked by measuring in each of the three axes separately. The uncertainty of 2D length measurement is checked by measuring in the XY, XZ, and YZ coordinate planes at a 45° angle to the corresponding coordinate axis. The uncertainty of 3D length measurement is checked by measuring along the operational cube of the machine or along the diagonal of a rectangular parallelepiped at a 55° angle to the corresponding coordinate axis. All coordinate measurement systems take part in the measurement process.

For each of 105 measurements (five different lengths, each measured three times in any seven different configurations, chosen at the discretion of the user) the error of length measurement is calculated, being the difference between the CMM readings and the true value of the corresponding length. The systematic error of the value of readings can be corrected at a certain position and direction if the CMM has a device (or software) for correcting that systematic error. Following the manufacturer's recommendations relating to environmental conditions, no manual corrections are allowed for estimating the effect of correctional calculations on temperature deformations or other factors. The true length standard value is considered to be the calibrated length between the end surfaces of the material standard. This value can be corrected for temperature changes only if this is included in the software of the test CMM. The error of the length being measured (as well as that of the displacement of machine parts) is equal to the difference of the machine readings in a certain coordinate and the readings of the base length standard.

The verification of a CMM is confirmed if not one of the 105 error measurements (expressed in micrometers) exceeds the E value specified by the manufacturer. A maximum of 5 out of 35 test length measurements can have one out of three repeated length error measurements that exceed the value of E. Each of these measurements must then be repeated ten times. Repeated verifications are performed following the same rules.

5.2 PRINCIPLES OF MHs AND TOUCH TRIGGER PROBES

The final part of the CMM is the MH that is usually attached to the ram of the machine. MHs vary in many ways, both constructional and functional. The most commonly used probes on the CMM are contact or touch–trigger probes. A touch–trigger probe contacts a surface on a part and generates an electronic signal to record its dimension. Noncontact probes are also used for scanning modes of measurement, that is, laser scanning systems, and so on. They are used to measure small, flexible parts or for measurement of parts made from soft materials where the surface may not be touched by any hard contact.

Touch–trigger probes are ideal for inspection of 3D prismatic parts and known geometries. These probes are highly versatile and are suitable for a diverse range of applications, materials, and surfaces. A wide range of accessories are also available for them. A hard probe, used for manual inspection of parts, is a solid contact probe consisting of a precision tapered shape tip (stylus) ending in a wear resistant ruby ball, usually a synthetic ruby. The probes are segregated into two categories, probes without, and probes with stylus module changing. Stylus module changing is a very important consideration as it enables higher productivity and the ability to always select the best measurement solution for the application.

An electronic trigger signal is generated by the probe using the contacts, as it is shown in Figure 5.2. During the contact with the surface point, the stylus of the probe deflects from the central position and contact is interrupted, thus generating a signal for coordinate fixing. Such sensor functions are also performed by capacitance, inductive, photoelectric, or piezoelectric sensors. The probe is very sensitive and of high accuracy, reaching about 0.1 μm of repeatability and uncertainty. Before measurement the home position is identified as the central point in the coordinate system of the machine and its value is set to zero. Home position is also referred to as the origin of measurement and is determined using a reference ball or rectangular body fixed to the table of the machine.

There are many constructions of MHs or probes that are equipped with rotation facilities on predetermined angles, having a large set of extension and stylus configurations as shown in Figures 5.3 and 5.4. This brings an additional axis of inspection capability by optimizing the working volume for the measurement. The probe head offers additional measurement capabilities and associated time savings. Usually, there are facilities for rapid head orientation and probe position determination.

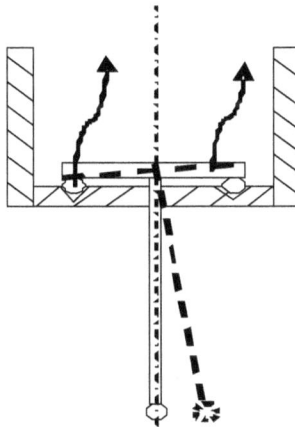

Figure 5.2. Touch–trigger probe with contact sensor.

Figure 5.3. Renishaw's PH20 touch-trigger probe CMM inspection head having facilities to rotate the stylus into the required position.

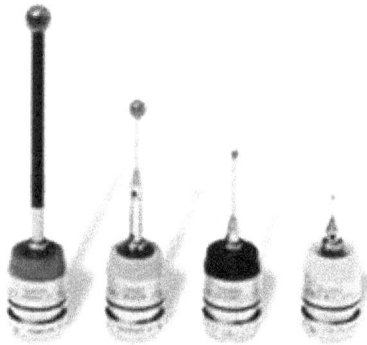

Figure 5.4. Different sizes and configurations of Renishaw MHs.

5.3 PERFORMANCE VERIFICATION OF MHs AND TOUCH TRIGGER PROBES

The final element in the total calibration chain of the CMM is the MH as the errors of touching the surface of the part add to the total error balance of the CMM.

Probing system testing consists of using the probe (MH) of the CMM to be tested and a material reference spherical artifact (a spherical body) as a substrate. The diameter of the spherical body is between 10 and 50 mm, the most used diameter being 30 mm, and the calibrated error of its total run-out should not exceed the value R (R—probing error). It is recommended to take at least 25 random point measurements on the surface of the reference sphere, approximately evenly distributed on the surface. Applying the least squares method, the center of the sphere is calculated and it is compared with the value stated by the manufacturer. All these procedures could also be applied to every CNC machine calibration, the only difference being a cutting instrument instead of an MH at the end of the calibration chain. Nevertheless, a disadvantage of the above methods is that the MH is calibrated by touching only a few points on the surface of its tip, leaving a large number of points without checking.

A special artifact has been developed for MH calibration (Figure 5.5). Instead of the widely used standard ring or convex ball a concave surface of very high accuracy and quality was used. A rectangular body of quartz glass was designed with a 60 mm diameter concave hemispherical surface

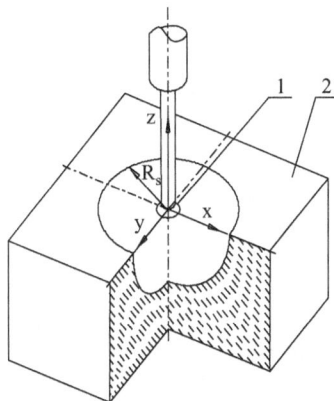

Figure 5.5. A quartz glass artifact for the calibration of the measuring head: (1) the tip of the measuring head; (2) a glass artifact.

in it. The dimensional accuracy of such a hemisphere is within ±0.05 μm in every section.

All sides of the artifact were lapped to the highest quality and accuracy. The radius of the hemisphere is $R_s = 30$ mm, a typical radius of the tip of the MH is 2.5 mm. The standard deviation of flatness, rectangularity, and circularity in all cross sections of the sphere is within the range ±0.02 μm to ±0.05 μm. The same range of tolerances for sphericity is assumed to be valid. During the calibration the tip of the touch probe is placed at different heights inside the inner surface of the concave sphere. Length end bars are successfully used for changing the height of the position of this artifact. It is a great advantage to have the possibility to move the tip of the MH only by several steps in the vertical direction and then to perform movements only by the machine drive connected to the glass artifact. The distance of displacement is measured by an LI or the measuring system of the CMM, if it is already calibrated. The distance measured is compared with that calculated according to the geometrical expressions determining the spherical surface of the artifact.

An advantage of such a calibration method is that the machine or special instrument used for the MH calibration must only perform 4 strokes, instead of the usual 12 strokes during the conventional methods of calibration using a convex spherical artifact. The strokes of the machine needed for the calibration of the MH using convex (12 strokes) and concave spherical artifact (4 strokes) are illustrated in Figure 5.6.

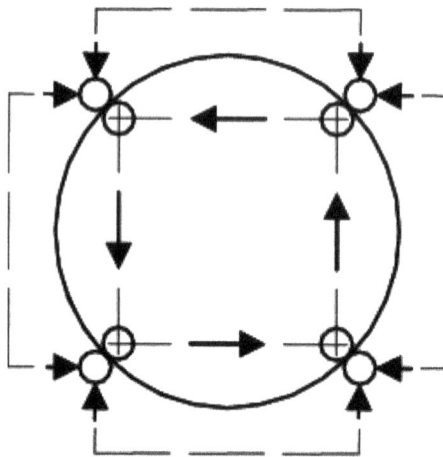

Figure 5.6. The strokes for the calibration of the measuring head using the inner or outer hemispherical body artifact.

The calibration results are processed as usual, calculating the scatter of touch point displacement values in terms of their difference from the sphericity of the artifact's surface. Then the positioning errors of all the elements of the kinematic chain of the CMM are known.

The results of the calibration are usually applied for compensation of the errors determined. The position of the final point of the measuring probe has a very important influence on the measurement in the case of a CMM. The correction of the final point of the tip of the MH can be performed by a correctional device inserted into the MH. Piezoelectric correctional devices can be effectively used for this purpose. The possibilities of piezoactive (piezoelectric or magnetostrictive) materials have rapidly increased, and their potential as precision actuators for correctional displacements are now widely realized. Recent developments of new active materials with extremely high piezoelectric or magnetostrictive properties (Terfenol-D, PZT, lead magnesium niobate, flexible piezoactive materials, etc.) have extended the area of their applications, including new systems with a high level of integration and multifunctionality.

5.4   COMPLEX ACCURACY VERIFICATION OF CMMs

The manufacturer is responsible for all technical parameters of the equipment and its compliance for the work intended. The equipment must be properly transported to the workplace and installed. The environmental conditions must also comply with the requirements set by the manufacturer. The equipment answers the requirements when the work tests included in the technical specifications are performed and are evaluated positively. Some of the working tests can be omitted according to mutual agreement between the supplier and the user. Working tests, such as repeatability, accuracy of linear displacement, working volume, and double length measurements can be carried out using various methods. These methods can show different results when they are performed on machines of different types. The results depend on slight differences of measurements and on the object of measurement.

Two main steps are used for the accuracy control of multicoordinate machines, robots, and measuring equipment. The first step is to calibrate the accuracy of the geometric elements of the machine; the straightness, squareness, flatness of guideways and bedding; the bias of vertical, horizontal, and longitudinal traverse, pitch, roll, yaw, and so on. The second step consists of measuring a reference part which has been previously calibrated. The results of the measurement of such a part on the CMM are compared with the accuracy results of this part collected during its calibration. Geometric features are assessed while aligning the coordinate system during calibration by the manufacturer and later by the customer's quality control staff. The alignment of the machine's coordinate system and the correction of the systematic errors are mostly dependent on geometric errors and their distribution characteristics.

The errors listed above must be controlled by the methods specified in the published standards or according to the standards or technical specifications supplied by the manufacturer. Special accuracy requirements are required for the MH of a CMM. The large number of accuracy parameters, methods, and instruments makes the task of calibrating multicoordinate machines very difficult, complicated, and expensive. The two main problems to be solved during coordinate measurement are to gather information about the errors of the object and to reduce the number of points to be measured to a minimum. This is even more important when using CMMs or during the calibration of the CMM itself for error determination in the multidimensional volume. The CMM constitutes a very significant part of the quality control system regulated by ISO 9001 [28]. Every measurement of geometric features on the part is to be validated by traceability, it must be related to stated references, usually national standards and stating uncertainties in this comparison chain.

As specified in ISO 14405 [29,30], the calibration is a set of operations that establish, under specified conditions, the relationship between values of quantities indicated by a measuring instrument or measuring system, or values represented by a material measure or a reference material, and the corresponding values realized by standards. Periodical verification of the CMM must also be performed, ascertaining and confirming the fact that the measuring equipment complies with the statutory requirements.

The CMM is calibrated according to its geometric accuracy parameters that are common with all metal cutting tools and machines and are presented and discussed in Chapter 2. The MH is to be calibrated according to the same accuracy parameters. During a full geometrical calibration, 21 detailed parameters are determined. However, other general testing methods are based on the application of special test bodies called standards or artifacts. Such testing is simpler, faster, and cheaper compared with measurements of all the error sources, which is costly and time consuming. Methods of measurement of the accuracy parameters of CMMs are shown in Table 5.1 below.

The first problem to be solved in the development of CMM performance verification methods is to gather a large quantity of accuracy data

Table 5.1. Methods used to measure accuracy parameters of CMMs

Axes	Types of error	Standards used for measurement
Linear	Positional errors	Laser interferometer, step gauges, linear scale, or translational transducer
	Straightness errors	Straight edges and wires, straightness interferometers, alignment lasers
	Squareness errors	Mechanical embodiments of the 90°, optical squareness standards, length standards inclined with respect to the axes of movement, rotary encoders
Rotational	Angular position	Standard rotary encoders, precision polygons, circular scales
	Pitch and yaw errors	Differential interferometers, autocollimators, electronic level meters, paraxial length measurements in several measurement lines
	Roll errors	Electronic level meters in horizontal axes, flatness standards, straight edges whose parallel displacement is monitored
	Camming	Test cylinders, flatness standards
	Axial rest	Test balls, flatness standards
	Run-out	Test cylinders, test balls

in a short period of testing, using simple reference equipment of low cost. The second step would be the efficient use of data selected for the accuracy improvement of the machine.

Various measurement models are used for sampling strategies during the selection of the optimal number of points in the measuring volume. The methods used are comprehensively analyzed in several research works and are discussed in more detail in Chapter 1. Sequence models such as those of Hamersley [8,9] have been used to calculate the coordinates of measurement points according to particular formulae for flatness measurement. Other authors have used the gray theory to predict measurement points, B-splines, and other methods of parameter modeling. When analyzing 3D space and the law of flatness error distribution, the various mathematical models used allow for the evaluation of the errors in the measurement space or plane. The most analyzed methods are associated with the set of mathematical models, generated series or splines.

The coordinates of measurement errors often depend on the type of spline, and if it is changed, another set of coordinates is generated. Moreover, the coordinates for error measurement depend on the model chosen for dividing the space or the plain into pitches (steps) for measurement. On an area where errors vary evenly the pitch of discretization can be larger, and on areas where the variation of errors is rather sharp the pitch of discretization should be significantly less.

Some other errors of the environment (temperature, pressure, etc.) are acting outside the machine and are analyzed and assessed in some other investigations. Specific errors are present in the signal formation of the MH of the CMM that must also be checked using specific test methods.

The problem of complex accuracy control of machines is under constant development and different methods for their resolution have been proposed. The simplest one was the use of a standard linear scale positioned in the diagonal direction on the working surface of the machine. Such measurements are to be completed on two levels in the working volume and they permit the selection of error information due to movement of parts of the machine. The large quantity of information in the measuring volume and lack of a spatial reference measure are the main technical and metrological tasks to overcome.

The complex measurement of the accuracy parameters of an automatic machine can also be performed by using the same kind of machine as a master or reference machine, both controlled by a single control unit. The final position of the machine's element or cutting tool, or the measuring probe for the CMM, is compared with that of the master machine. All the accuracy parameters of the machine are involved in the determination of this position. An example of such a calibration process using an LI and fiber optic links is presented. The information from the laser unit and from the IMSs of the machine or CMM is delivered to a computer where the results of measurement are processed and presented in the desired form.

Figure 5.7 shows the two machines, (a) is the master (reference) machine and (b) is the machine to be calibrated. There are identical moving parts on the machines, the bridge moving in the y axis direction, the carriage mounted on the bridge moving in the x axis direction and inside the carriage, the ram moving in the z axis direction. The LI is fixed to the carriage of the master machine and moves together with it along the axis x. On the path of the light beam of the laser there are reflecting mirrors (RM1, RM2, RM3, RM4, and RM5) and the semitransparent (semireflecting) plates S1, S2, S3. The optical path of the LI between the two machines is linked by fiber optic cables (FOC).

(a)   (b)

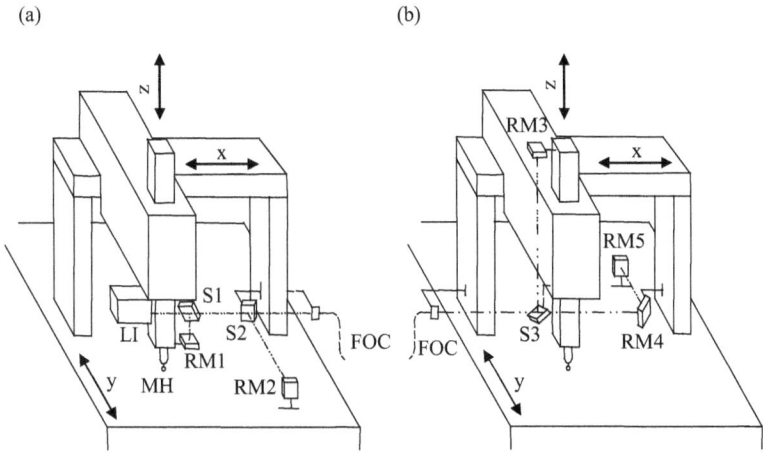

Figure 5.7. Calibration of the accuracy parameters of the CMM (b) against the reference (master) machine (a).

The characteristic feature of the calibration is the path of the laser beam going through all the coordinate movements of both machines. It is important that the reflecting mirrors be fixed on the opposite sides of the strokes of movable parts. In this case, the length of the laser light path remains unchanged when the coordinate distance of the parts (bridge, carriage, or spindle) of the machine to be calibrated and that of the other (reference) is changing. Only systematic errors of coordinate movements are registered in the computer. Environment conditions of the control also have a minor influence, as the calibration is performed under the same conditions for both machines. The method proposed enables the operator to obtain quite a large amount of accuracy information and to use it for the accuracy assessment of the machine and provide error compensation as well.

Denoting the coordinate displacements of the master machine and of the machine to be measured as follows: Δx_i, Δy_i, Δz_i and $\Delta x_i'$, $\Delta y_i'$, $\Delta z_i'$, and the lengths of the strokes l_1, l_2, l_3, the error of the coordinate movement will be expressed as

$$\delta_{x,y,z} = \left(\sum_{i=0}^{l_1} \Delta x_i' + \sum_{i=0}^{l_2} \Delta y_i' + \sum_{i=0}^{l_3} \Delta z_i' \right) - \left(\sum_{i=0}^{l_1} \Delta x_i + \sum_{i=0}^{l_2} \Delta y_i + \sum_{i=0}^{l_3} \Delta z_i \right). \quad (5.1)$$

The mirrors RM1 and RM3 are fixed to the rams of the two machines, the mirrors RM2 and RM5 are fitted to the bases (tables) of the two machines.

RM4 is an angular reflecting mirror and it is fixed to the bridge structure of the machine to be measured. Semireflecting plates S1 and S3 are fixed to the carriages of the machines, and the semi-reflecting plate S2 is fixed to the bridge of the reference machine.

Diagrams of the arrangement of the LI with the reference and measuring CMMs together with the path of the laser beam between the two machines are shown in Figures 5.8 and 5.9.

The information from the laser unit and from the measuring systems of the machines is fed into a computer where the results of measurement are processed and presented in a desirable form for evaluation. The characteristic feature of the control is the fact that the path of the laser beam goes through all the coordinate movements along the axes X, Y, and Z, of both machines. Also it is important that the reflecting mirrors RM1 to RM3; RM2 to RM5 should be fixed on the opposite sides of the strokes of movable parts. In this case, the length of the laser light path remains unchanged when the coordinate distance of one machine is enlarged, and that of the other is shortened accordingly. In all cases only errors of the machine parts' strokes are registered in the computer. Trials have shown that it is easy to compare the strokes of the machines. Of course, this method can only be used by companies manufacturing such machines or

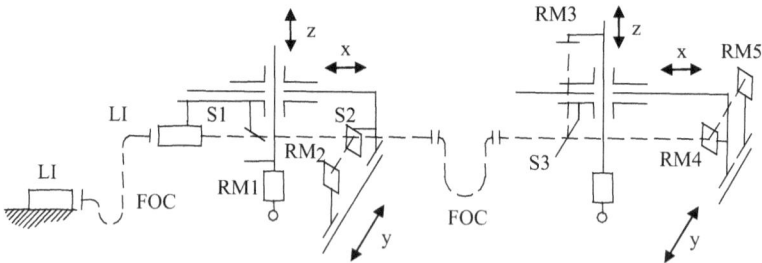

Figure 5.8. Calibration of CMM coordinate positioning accuracy against the master machine.

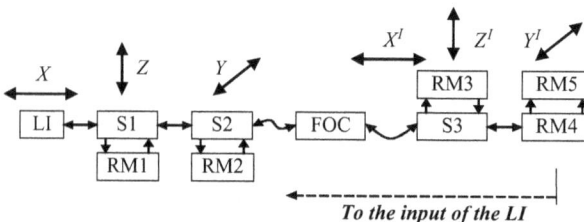

Figure 5.9. The diagram of the path of the laser beam.

with sufficiently strong resources to possess a number of similar machines. The environmental conditions should not affect the calibration results as the calibration is performed under the same conditions for the both machines. This enhances the accuracy and productivity of the calibration process.

5.5 PERFORMANCE VERIFICATION USING ARTIFACTS

Mechanical artifacts are reference standards (objects) used for CMM performance and accuracy verification. They are used to compare measurements performed on them using the CMM with previous results of measurement of this artifact. Hence, the functional capabilities of the CMM are verified, adding the accuracy biases received by measurement. Traceability of measurements is also ensured as prior calibration of the artifact is linked to the length or angle standards.

Artifacts are usually made from thermostable materials or granites and have several numbers of spatial coordinates associated with their principal calibrated features. Combinations of flat surfaces, holes, balls, inner or outer spherical surfaces, rectangular surfaces, and other geometrical features are commonly used in the construction of artifacts. An example of a granite artifact containing holes and balls is shown in Figure 5.10. This artifact, in common with many others, can be placed in different axial directions or aligned at a predetermined angle according to the CMM's coordinate axes.

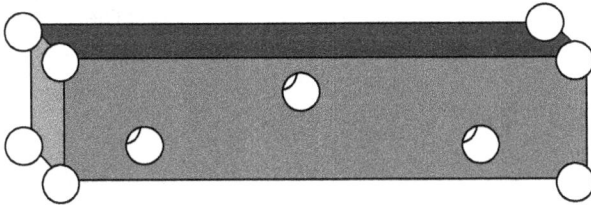

Figure 5.10. Example of granite artifact containing different geometrical features.

An example of a granite artifact with the holes placed according to the *L-P* sequence is shown in Figure 5.11 (see the section on *L-P* sequences in Chapter 2).

To enable the artifacts to be more easily handled on the table of the CMM, they are often made from a light frame construction to which balls, holes, or other geometric features are fixed at some convenient points.

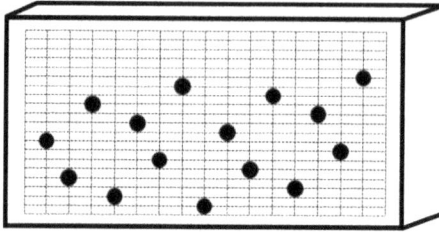

Figure 5.11. Example of granite artifact with the holes placed according to *L-P* sequence.

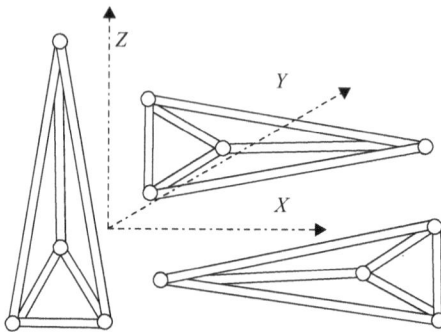

Figure 5.12. Four-ball space frame.

An example of this kind of artifact, a four ball space frame, is shown in Figure 5.12.

Artifacts are one, two, or 3D reference measures. The most commonly used types of artifacts can be placed into the following groups:

• length bars,
• gauge blocks and step gauges,
• ball-end bars,
• ball and hole plates,
• straight edges and squares,
• space frames.

One of the simplest test methods uses an artifact made from gauge blocks (Figure 5.13). The test consists of measuring length standards in various positions and orientations in the working volume of the machine. Such an artifact has an accurately known length between its two flat and parallel end faces. ISO 10360: Part 2: 2009[14] specifies five different lengths for

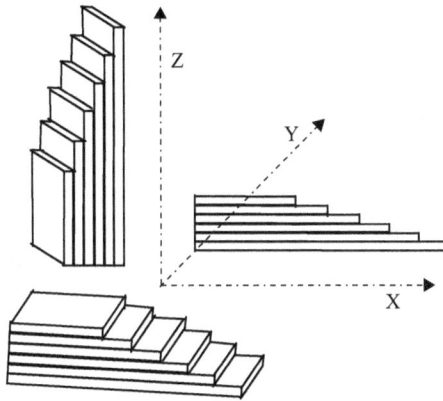

Figure 5.13. Artifact made from a set of length gauges.

measurement in seven random positions, taking three measurements for each length in each position.

Measurement can be performed using end-length measures and an MH. At the first point of contact by the end-length measure the readings of the machine must be set to zero. During each travel three measurements must be performed and an average arithmetic result is calculated. The process is repeated in all three coordinate axes. To minimize the influence of environmental conditions all measurements should be made in as short a time period as possible.

To determine the measurement uncertainty of 1D, 2D, and 3D measurements, one-, two-, and 3D measurements are employed. They differ in the number of directions in which displacement is performed, and the corresponding coordinate measurement systems which are at work during measurements. The uncertainty of 1D length measurement is checked by measuring in each of the three axes separately. The uncertainty of 2D length measurement is checked by measuring in XY, XZ, and YZ coordinate planes at a 45° angle to the corresponding coordinate axis. The uncertainty of 3D length measurement is checked by measuring along the operational cube of the machine or along the diagonal of a rectangular parallelepiped at the 55° angle to the corresponding coordinate axis. All coordinate measurement systems take part in the measurement process.

For each of 105 measurements (five different lengths in seven random positions, taking three measurements for each length in each position) the error of length measurement is calculated, that is, the difference between the CMM readings and the true value of the corresponding length.

The systematic error of the value of the readings can be corrected at a certain position and direction if the CMM has a device or software for correcting for the systematic error. Following the manufacturer's recommendations regarding environmental conditions, no manual corrections are allowed for estimating the effect of correctional calculations on temperature deformations or other factors. The true length standard value is considered to be the calibrated length between the end surfaces of the material standard. This value can be corrected for temperature changes only if such a possibility is provided by the software of the test CMM. The error of the length being measured (as well as that of the displacement of machine parts) is equal to the difference of the machine readings in a certain coordinate and the readings of the base length standard. From the measurement results two graphs are drawn, and the error of linear positioning is determined either as the maximum difference between the graphs or the positioning error of a certain portion of travel.

5.6 PERFORMANCE VERIFICATION USING LASERS

LIs are used for length measurements and also for pitch and yaw, parallelism, out-of-squareness between two perpendicular axes, positioning accuracy, and straightness measurements. These operations are described in Chapter 2. These accuracy tests are performed according to BS 3800: Part 2: 1991 [31]. Manufacturers and users of CMMs have all used LIs for many years, mainly during the build and on-site servicing of their machines. The LI is a powerful tool for many accuracy tests. Accuracy test procedures are axis-by-axis proof of the accuracy build-up of the machine and are not considered as performance verification of the machine. Such parametric calibration tests are to be performed by users at regular intervals during the working life of the machine.

Some measurement patterns are used for assessing the positional error using an interferometer system coupled with a swivel mirror. This set enables the accessibility of the laser beam in four cross-diagonals. The diagram of measurement is shown in Figure 5.14.

Some regulations must be complied for carrying out the performance test. The first is the arrangement of the measuring lines within the working volume of the CMM according to BS6808: Part 3: 1989. The second one is the alignment of the measuring line of calibration. The arrangements of the measuring lines are shown in Figure 5.15. Figure 5.15(a) shows cross-diagonal configurations, Figure 5.15 (c) shows in-plane diagonal

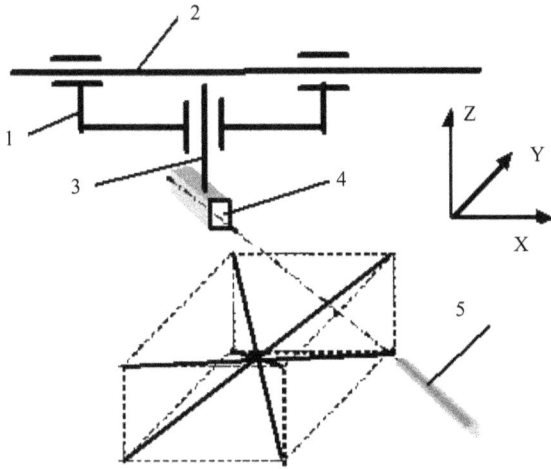

Figure 5.14. Diagram of measurement by laser interferometer. (1) Machine carriage, (2) slideway, (3) ram, (4) swivel mirror fixed to the ram of the machine, (5) laser beam aligned with mirror fixed to the ram.

(a) Cross-diagonal configuration

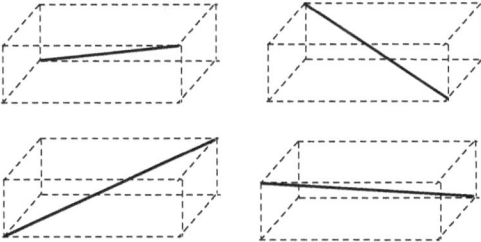

(b) Suggested additional measuring line

(c) In-plane diagonal configurations

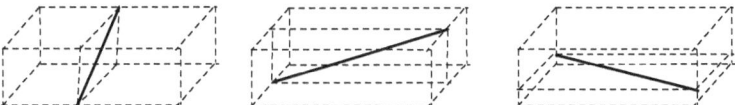

Figure 5.15. Orientation of measuring lines.

configurations, and Figure 5.15 (b) shows a suggested additional measurement line. These length measurements can be performed by using both length gauges and an LI.

Gauge blocks can be available in the form of step gauges. The step gauge can be made from a single piece of the basic material or it can consist of gauge blocks inserted in the basic body. The gauge blocks allow external measurements and the step gauge permits not only external but internal and step measurements of different sizes in multiple positions along a common measuring line. More measurement possibilities are provided using the LI. However, the alignment of the laser beam and the mirror fixed on the ram is a more complicated task and takes more time in volumetric measurements.

5.7 METHODS AND MEANS FOR ACCURACY IMPROVEMENT OF MACHINES

The development of mechatronic devices for the accuracy improvement of measuring and metal cutting machines, robots, and so on, provides a great opportunity for accuracy improvement of machines after their calibration. Mechatronic methods for the correction of systematic errors of information-measuring systems, including linear and rotary encoders and final 3D position correction of a cutting instrument or touch probe, are presented. Technical solutions are provided for the application of active materials for smart probes and piezoactuators. The application of piezoactive raster plates and other elements is described together with volumetric control diagrams and examples of practical implementation. Error correction problems can be solved using a piezoactive, piezoelectric, or piezomagnetic substrate in the measuring or the datum system of the machine. The implementation of piezoelectric correctional devices is very efficient for precise engineering, where cutting or measuring forces are negligible and displacement values are in the range of several micrometers.

The play of the bearings, the change of the position of the moving parts due to their weight, wear, environmental influences, and so on, reduce the final accuracy of the machine. Existing methods for increasing the accuracy of precise equipment are generally complicated and of high cost. A new concept in the design of error compensating systems is presented here which includes active frames and structures for precision technological equipment. The most effective way to improve the accuracy of machines is to control and correct a position of the final member of the kinematical chain of the machine. Such an element could be the tip of a cutting instrument or the tip of an MH (touch probe) of a CMM. Piezoelectric or other types of transducers can be applied for the final correction of the position of an element in a 3D machine. Equations for the calculation of correctional displacement are derived and possible fields of application are reported in several publications. Crawley and De Luis [32] proposed the insertion of active elements (e.g., piezoceramic or TERFENOL-D plates or components) into the structure of the machine and controlled their displacements with electric or magnetic fields applied in the appropriate value and place. It has been shown that by implanting active links into the structure of the equipment both static and dynamic displacements can be sufficiently controlled.

The application of electric or magnetic fields generates displacements of piezoplates to compensate for the systematic errors of a scale or of the elements of an encoder. The scale is used as the main element of the optical measuring system of a machine or it serves as the main part of the translational (rotational) transducer in a measuring system, so the correction of the systematic error of measuring scales is quite an important and effective method for accuracy improvement. Some considerations for a new approach to error compensation circuits and actuators are presented here. The compensation covers not only conventional measurement systems, but also opens new ways of increasing the accuracy of typical components of technological machines, such as beddings, slideways, fixing and datum surfaces. New structure units for precise correction mechanisms—*active kinematic pairs*—were presented by Bansevičius and Giniotisin [33]. They can be applied to photoelectric or other types of transducers and for the final correction of 3D machine element positions. Equations for the calculation of correctional displacement were derived, and possible areas of application shown. The integration of the unique properties of piezoactive transducers and actuators (high resolution, low time constant, easy control of forms, types and parameters of oscillations, possibility to generate multicomponent static, quasi-static and resonant displacements, high electromechanical conversion efficiency) with the control system made it possible to reduce significantly or even fully eliminate the main errors of bearings, supports, and guides used in high precision measuring devices and machines. The number of constraint conditions can be varied in different ways. The simplest is the control of the friction, acting between a pair of elements, usually when the elements of the pair are held together by force. Here either the friction coefficient or the magnitude of the force executing the closure can be changed. This is achieved by the excitation of high frequency tangential or normal oscillations in the contact zone of the pair. Electro rheological and magneto rheological fluids whose viscosity can be varied within a wide range can also be used successfully. Elements of piezoactuators are produced in a great variety of shapes, such as squares, rectangles, rings, discs, spheres, bars, and cylinders. The piezoelectric effect is linearly dependent on the applied electric field and such a technique can be used to great advantage in precision engineering.

5.7.1 PIEZOMECHANICAL CORRECTION

Design of correction mechanisms normally involves rather complicated mechanical, electromechanical, or programmable systems. These systems are not universally applicable due to considerable size, complexity of large transmission ratios of backlash-free mechanisms, or devices for coordination of displacement with a programmable correction system. These problems are especially critical in cases when there is a need to increase the accuracy of a measurement system of small size, built into vacuum chambers and devices subjected to various external or temperature, or both, disturbances.

By connecting an electric supply of different values to the different parts of apiezoplate, it is possible to change the deformation of the plate, and thus perform linear systematic error correction. The problems of linear information/measuring system correction can be solved using a raster scale on the piezoactive, piezoelectric, or piezomagnetic substrate. In such designs, an electric or magnetic field generates the displacements that compensate for the errors of the shape or the pitch of the raster scale. Photoelectric raster transducers as well as raster scales are equipped with piezoelectric or piezomagnetic parts. An electric excitation is connected along the length of the raster scale to be corrected which will cause microdisplacement of the raster pitches, compensating for the error determined. The pitch error control must be calculated beforehand and the longitudinal displacements are related to deformations by the piezoelectric plate. The main equations for such calculations can be used to determine the voltage supplied to the piezoelectric grating in the interval of the correction. Equations are derived for the one and two-dimensional cases by application of the finite element method. The coefficient of electromechanical coupling, the size of the electrode coating of the piezoplate, modulus of stiffness at constant electric displacement, the electric field in the direction of displacement, and the piezoelectric constant are taken into account when deriving the equations.

As an example the former pitch error of a raster scale was improved by applying different voltages from a DC source to different sectors of piezoelectric plates attached to the raster scale or to the plate inserted between the indexing head and the moving part of the machine. The applied voltage ranged between 200 and 500 V but rapid development of piezo-active materials provides an opportunity to perform such tasks with lower voltages and smaller dimensions of parts in the correction mechanism. Overall correction can be accomplished by using piezoelectric plates from one or the other end of the transducer. The error graphs represent the pitch error in the former state and after the correctional activation. The systematic error can be reduced to more than half its former value. It is obvious that the correction system becomes more compact. The correction can be performed for the systematic error and to the periodical (high frequency) error in designated segments of the raster scale or the translational transducer. The correction is effective for reducing temperature errors and to correct installation errors of measurement systems. The correction is possible both for linear and circular raster scales and transducers.

Every complicated measuring system belonging to the field of mechatronics has a mechanical structure, including a drive controlled by a PC or microprocessor, and so on. It may not only be a one DOF micro-displacement or nano-displacement system, but it may also be a multidegree of freedom displacement system. Research has been undertaken on the accuracy of multiparameter systems and the solutions for determination of the deviations of those parameters for the purpose of application to adaptive correction and for assurance of mechatronic control of the system.

Industrial equipment with its high technology measurement and communication systems deals with tremendous volumes of information data and their associated input/output complexity. Modern measuring systems consist of "smart" transducers that can perform some logic functions and "smart" sensors and actuators, together with complicated control systems with special programming languages able to control the functions of a measuring system by digital and alphanumeric information.

Many angular and linear measurement transducers or encoders are used in industry and machine engineering for position and displacement measurement. The accuracy of angular position fixed by means of these devices reaches approximately $0.3''-0.1''$ (seconds of arc) and the amount of data in such systems sometimes exceeds gigabytes. Information about the quantitative and qualitative state of an object is selected using methods of mathematical statistics. The information selected and processed is used to determine a statistical estimate of the mechatronic object that is required for its essential apprehension and for relevant impact, for example, making error correction [27,33]. This includes information about the technical parameters of the object to be controlled, its displacement with the accuracy parameters and the control of the position and micro/nano displacement. It is especially important for the position accuracy control in nano displacement systems, information-measuring systems, such as raster and coded scales of high accuracy, transducers, rotary encoders, and so on.

The measurement intervals for accuracy calibration of length and angle displacements in machines are given in relevant technical documentation. These requirements stated by written standards show the same problem, that is, information inside the given interval of measurement remains unknown. This problem is analyzed in several publications and measurement information is supplemented by evaluating the information and joining it with the general expression of the measurement result, that is, expressing the result in terms of the systematic error, the uncertainty of the assessment, and the probability.

$$X = \bar{x} \pm u(x_i), P,$$

where X is the result of measurement; $\bar{x}$ is the systematic error; $u(x_i)$ is the standard uncertainty of measurement, and P is the probability.

This means that the measurement result would be more informative, adding to it a parameter that shows the extent of the sample assessed during the measurement process determined with the uncertainty assessed by probability level P and with the indeterminacy of the result assessed by estimating the portion of the sample from all the data of the IMS.

Mutual information is a measure of the amount of information that one random variable contains about another random variable. In other words, it is the reduction in the uncertainty of one random variable because of the knowledge about the other. Therefore, it is important to determine the information quantity on an object that has been evaluated providing a more thorough result assessment during the accuracy calibration processes. The problem exists due to the great amount of information that is gained in the calibration of scales, information-measuring systems of numerically controlled machines, and automated measuring equipment such as CMMs in their total volume. It is technically difficult and time consuming to calibrate the enormous number of points available, for example, the 324,000 steps of the rotary table or every discrete piece of information from the measuring systems of metal cutting tools and CNC machines, some of them utilizing LIs.

These assumptions can help in the determination of the mechatronic impact on the information-measuring system for the elimination of systematic errors. Piezoplates and simplified bodies constructed from piezo-material are applied for micrometric displacement in order to correct the systematic errors of raster scales. Correction of errors is accomplished by control of the potential energy of the piezoplate and this can be achieved by changing a voltage supplied to the electrode, electromechanically coupled with the raster scale, thus changing the width of the electrode coating of the piezoplate. According to the principle of minimum potential energy $\dfrac{\delta U}{\delta \varepsilon_1} = 0$, and taking into account that longitudinal displacements are related to deformations by the relationship, $\varepsilon_1(x) = \dfrac{du}{dx}$, it is possible to obtain the equation of equilibrium of the piezoplate considered and to control the accuracy of a linear raster scale. The distributions of errors for two-dimensional or 3D measuring systems can be expressed by two functions $u(x,y) = -k_1 \delta_x(x,y)$ and $v(x,y) = -k_2 \delta_y(x,y)$, where u and v are the components of displacements at the zone controlled in the direction of appropriate coordinates. k_1 and k_2 are the coefficients of electromechanical coupling.

A simple piezoelectric system is shown in Figure 5.16 for use in the centering device of a rotary table. It is useful for circular scales measurement and part machining on the rotary table. Two pins with piezoelectric tips are used for displacement of the inner ring with the part to be centered according to the basic part, the hub. Displacement information can be controlled manually or it can be arranged in an automated way when the system is supplied by displacement transducers or gauges.

Piezoactuator elements are easy to insert into some parts of this system. In practice these elements can be from 20 µm to 15 mm wide and up to 100 mm length or external diameter. The maximum possible relative change in length is up to 0.13 percent and will be either an expansion or contraction after applying or changing the voltage. After disconnection

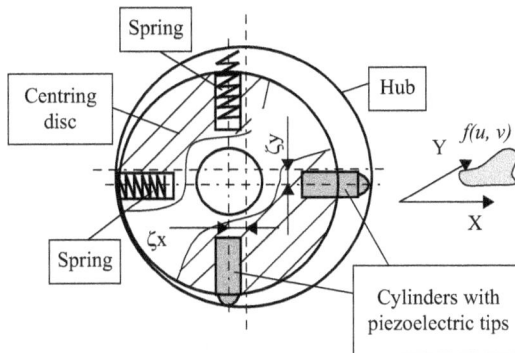

Figure 5.16. The diagram of a mechatronic centering device.

from the power supply an expansion/contraction of the piezoelement decreases slowly. This results in a current:

$$i = \frac{dQ}{dt} = C\frac{dV}{dt},$$ (5.2)

where Q is the charge, C is the capacity, and V is the voltage.

The equation of equilibrium of the piezoplate may be expressed as follows:

$$e_{31}b_1(x)E_3(x) = b(1-k_{31}^2)c_{11}^D\frac{du}{dx}, \qquad x \in [0,l],$$ (5.3)

where e_{31} is a piezoelectric constant, $E_3(x)$ is the electric field in the plate, $k_{31}^2 = e_{31}h_{31}$ is the coefficient of electromechanical coupling. Since $u(x) = -k\delta(x)$ is specified for correctional purposes, to satisfy this equation it is necessary to provide the required value of the expression $[e_{31}b_1(x)E_3(x)]$ at every point x along length l. This is possible by using two different approaches:

(i) By changing $E_3(x)$ along the piezoplate when $b_1 = b =$ const. $b_1(x)$ is the width of the electrode coating of the piezoplate. Technically this is performed by means of segmentation of the electrode coating of the piezoplate along the length l and by applying a different voltage to every segment. The continuous function of $E_3(x)$ is equal to

$$E_3(x) = -\frac{1}{e_{31}}c_{11}^D(1-k_{31}^2)\frac{d\delta(x)}{dx},$$ (5.4)

c_{11}^D is the modulus of stiffness at constant electric displacement. The minimum length of the segment necessary for satisfactory approximation must not exceed $T/2$, where T is the period of the higher harmonic of the expansion $E_3(x)$ into a Fourier series.

(ii) By changing the width of the electrode $b_1(x)$ along the length of the piezoplate at $E_3 =$ constant. In this case, there is no need for a large number of supply voltages and the number of segments is appreciably smaller than in the first case.

By means of these expressions, there are no difficulties to obtain the values of intensities of the electric fields providing the necessary displacements of the piezoplate.

The correctional diagrams for two typical information-measuring systems of a machine are shown in Figures 5.17 and 5.18.

The general system shown in Figure 5.17 includes a linear or rotary encoder (consisting of a measuring scale and a reading head), electronic control and display units, programmable software for the process control. It is evident that these are mainly optical and electronic devices which have no heavy masses or large dimensions and no evident forces or velocities are present here. So, this is the best place to apply any correctional input for the compensation of the systematic errors of the encoder and the machine as a whole. Using the information received from the upper level of the digital counter, the output information is corrected by the preselected correctional value from unit 4. It is a convenient and simple means for the correction of linear systematic error, such as depicted in Figure 5.17(b). Another possibility for correction of the systematic linear

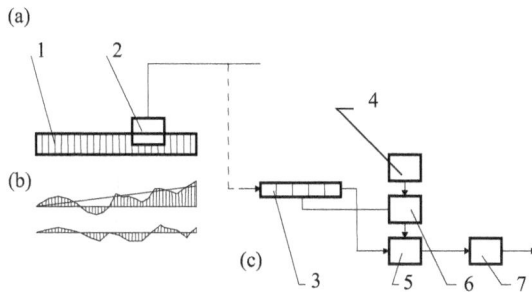

Figure 5.17. Correction of the information-measuring system of a machine. (a) translational transducer of the system: 1—scale, 2—reading head; (b) diagram of the systematic error; (c) the electronic diagram: 3—digital counter, 4—generator of the correctional value, 5—element OR, 6—conjuncture, 7—output unit.

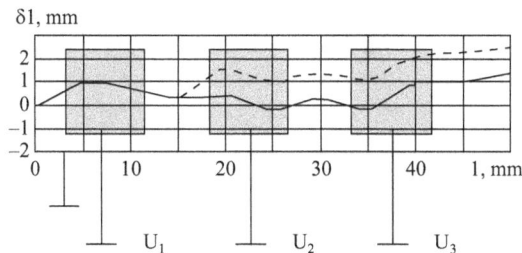

Figure 5.18. The correction of the systematic pitch error of the scale on the piezoelectric substrate.

error would be micro displacement of the reading head 2 supplied by a piezoelectric plate inserted between the head 2 and the construction of the machine, for example, the moving table of the machine.

A similar operation is performed by the device shown in Figure 5.18 where values of systematic error before correction (dashed line) and after correction (continuous line) are shown. The correction of errors over the whole length of the raster scale creates some difficulties. The scale of the encoder is supplied by the piezoplate with the electrodes placed along the length of the plate and connected to the voltages U_1, U_2, U_3, and so on. Values of the supply voltage are calculated according to equations (5.2–5.4). Experiments performed show an effective systematic error correction that improves accuracy about 1.5 times.

In more general cases, the implantation of active links into the technological chain of machines can compensate for unwanted displacements, caused by external forces or a change of the position of moving components, imbalance of rotating systems, wear of contacting surfaces, temperature influence, and so on. It is common practice to create drives and transducers capable of performing and controlling the measurement in two- or 3D directions. In this case, piezoelectric materials would provide the possibility to correct such displacements using only one piezoelectric body. The components of elastic stress and deformation in the 3D directions can be controlled by applying an electric field to the piezoelectric actuator. Measurement systems using piezoelectric correctional devices are more compact and a correction can be performed for both systematic errors and errors of small-period in the designated segments of the measuring length.

Using such correctional systems, some accuracy improvement problems of machines and measuring equipment can be solved. The accuracy of the coordinate measuring system of a machine is improved by mounting additional components into the structure, for example, piezoceramic plates or rings, allowing the control of small displacements in specific directions. Figure 5.19 shows the use of piezoelectric plates inserted in the relevant parts of the machine's slideways in the different axes of displacement, which can effectively facilitate for the correction of the main systematic errors. Displacements are controlled by applying an electric field in the same direction as that of the poling vector. The main task of the correction is to determine the correctional displacement values in all coordinate displacements by calculations and then to perform the additional movement using the last (conclusive) part of kinematic chain of the machine. For example, it may be the grip of the arm of an industrial robot, the touch probe of a CMM (measuring robot), the cutting instrument of a metal cutting tool, and so on. In this case, piezoelectric plates, cylindrical or spherical components are useful to incorporate into the appropriate machine member for the purpose of accomplishing the correctional micro displacement required. The final element of the CMM, a touchprobe (overall sensitivity about 0.2 μm) with the piezoelectric cylinder and the electrodes for electric supply are shown in Figure 5.19 (also see the section A-A). Piezoelectric active elements are also implanted into the sliding parts of the machine, that is, the console moving along the y axis, the carriage moving along the x axis, and the ram with the touch-probe moving in the direction of z axis.

One of the functions of a piezoelectric part of the MH is the formation of a signal due to the contact with the surface of the object to be measured. Another function is to move the axis of the MH by using the voltages U_1, U_2, and U_3 (section A-A) to generate the displacement necessary to correct the geometrical error in the relevant part of the measuring volume. The control of this displacement is performed by the computer in control of the CMM. When the input voltage is supplied to one, two, or three electrodes of the cylinder, due to the reverse piezoelectric effect, the cylindrical body

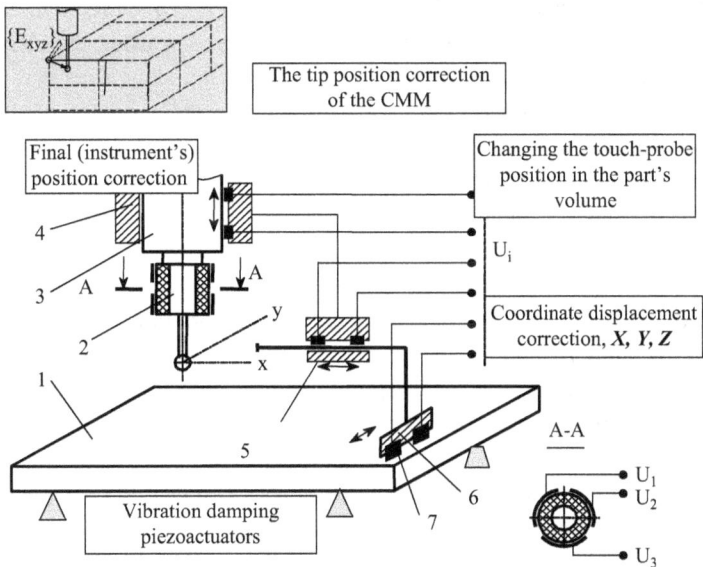

Figure 5.19. Piezoelectric correction of parts of the machine or the tip of the measuring instrument: (1) bedding, (2) touch probe, (3) ram, (4,5) carriage, (6) console, (7) piezoelectric plate insertions. U1, U2, U3—mains supply to the measuring probe; Ui—mains supply to the piezoelectric insertions.

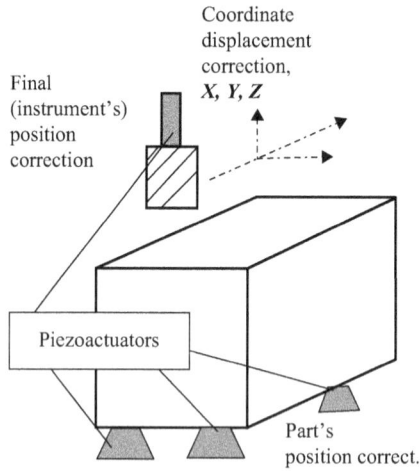

Figure 5.20. Correction of the position of the machine's displacement or the tip of the measuring (or cutting) instrument.

of the MH deforms and moves its axis into the desired position. The tip of the MH is displaced for the volumetric error $\{E_{x,y,z}\}$ correction. Such design permits multipurpose tasks of error correction using the electric supply U_i to the different parts of the frame (moving along axes x, y, and z on the bedding of the machine), so improving the accuracy of slideways, bedding, correcting the errors caused by moving parts, temperature changes, and so on. Such active elements due to their compactness have great advantages compared with many conventional correctional systems.

New active materials with high piezoelectric or magnetostrictive properties (Terfenol-D, flexible piezoactive materials, and so on.) show a wide area of application with a high level of integration and multifunctionality. They are applicable to implement into the piezoactive part's supports (Figure 5.20) giving an opportunity to correct the position of the part in 3D space, compensating for the systematic error of the machine's displacement.

The methods discussed in this section can control the accuracy of displacement of parts of a machine, of the transducers, or the final member of the kinematic chain of a machine, such as the touch-probe or the cutting tool. Piezoelectric active bearings for shaft or spindle accuracy control have also been designed and tested. Using these techniques geometrical error correction along the axes of a machine can be performed at a lower cost and more efficiently.

CHAPTER 6

FUTURE TRENDS IN MACHINE MONITORING AND PERFORMANCE VERIFICATION OF COORDINATE MEASURING MACHINES

Measurements of geometrical parameters of parts are normally performed on very expensive 3D measuring equipment, known as coordinate measuring machines (CMMs). Coordinate measuring machines have a high precision mechanical base equipped with very high precision slideways (often made of granite parts) of the highest accuracy requirements for straightness, perpendicularity, flatness, and so on. Quite expensive linear displacement transducers or rotary encoders are mounted into every axis of measurement. CMMs are an example of highly complex, precise and highly expensive equipment. Nevertheless, the very fast development of optical, radio, electronic, and mechatronic methods for sensing, tracking, data collection, transmission, and evaluation of data, allows a wide range of application of these techniques to be considered for many kinds of measurement. Some examples of the state-of-the art in this field and some proposals for the development and implementation of telemetric methods for measurement of the machined parts are presented in this chapter. The theoretical background for the implementation of a telemetric system for volumetric measurements is discussed.

6.1 INTRODUCTION

One of the major benefits to be derived from the successful outcome of recent developments will be that a heavy three-dimensional (3D) CMM construction with its associated motors and encoders will no longer be required to both accurately position the probe and to determine its exact location. Hence a parallel engineering development will investigate mechanisms for a lightweight, low-power drive system that can position a probe within 3D space. High accuracy of the basic parts of the machine is not necessary, as the accuracy of the system will be determined by the accuracy of the telemetric location system.

The aim of this new development is to eliminate very expensive mechanical and electromechanical basic parts in measuring devices by developing an information transfer system for the position, displacement, and dimensions of a machine part in the measuring volume. A measuring probe (MP), for example, an optical or photoelectric touch probe or non-contact probe, would follow the surface of the part to be measured, or a point on the moving object, using a transmitter–receiver technique (radio or optical waves) to send information to the fixed receivers and to the control unit for the evaluation of position or displacement. The reference measure is to be created using the triangulation points on which the receivers–transmitters are located. The measurement principles are valid for measuring and testing equipment and also for other technological equipment.

The first task for this implementation would be the development of new techniques for performance (accuracy) verification of CMMs in-service and the second task would be to create a new CMM, based on telemetric measuring systems, known as a telemetric measuring machine (TMM).

The design of a new type of instrument for CMM performance verification will improve the quality of measurement by eradicating the link between existing machine errors and the measuring devices and personnel errors. It will produce a significant saving in the amount of material, machining, and power consumed in the production and operation of a CMM. It could be expected that one technical achievement of this project would be to reduce the weight and volume of the verification devices by at least 50 percent. With the additional savings in machining and power requirements, this would bring considerable environmental benefits.

Special reference parts (artifacts) have been used for CMM calibration purposes until now. A variety of configurations of such artifact parts are used, which include rectangular plane surfaces, internal cavities or exterior spherical bodies, cylindrical and conical surfaces, and so on. The disadvantages of such methods of accuracy control are in the restrictions of the measurement volume of the machine to be assessed. It is inconvenient to handle a large dimension reference artifact on the table of the machine if such an artifact is used. Hence, most reference artifacts have small dimensions, so the calibration does not cover all the volume of the machine. In this case, a large volume inside the machine is unavailable for calibration. Consequently, some new methods and techniques for easier and more convenient calibration of the accuracy parameters of CMMs are necessary.

The expected financial savings to be gained from this development should help to reduce the comparable cost of CMMs for any given measurement range. Hence, the expected design benefits may not only make the CMM appropriate for a wider range of applications, but also the reduction in cost would make the use of a quality measurement device more readily available to other users (particularly small to medium sized enterprise [SME] companies).

The most advanced optical methods applied for 3D measurements, including laser tracking systems and sensors, together with the use of digital cameras for such systems are discussed in this chapter. They offer great advantages due to minimal dimensions, wide range of accessibility, and integration of the sensor and information transfer functions in one unit. It is expected that these modern systems will replace the existing 3D coordinate measuring systems, especially the ones using Cartesian coordinate systems, and they will become more widely used in the machine industry and in geodetic measurements. An example of the use of optical position determination by using optical bar marks is one of the possibilities for future development.

6.2 THE DETERMINATION OF PHYSICAL SURFACE COORDINATES USING A CCD CAMERA

The method of surface scanning using laser technologies is easy to realize from the organizational point of view but is more complex from the technical side. Such a method of surface scanning is accurate and point position standard deviation estimates reach a few micrometers [34–40]. However, other methods should be used if, due to the complex physical surfaces, the scanning ray cannot "see" some surface points. A method for the determination of physical surface coordinates using a CCD camera is presented here. A suitable scale is provided for the linear converter of photodiodes during CCD camera calibration. Point coordinates are determined by the method of linear intersection in a free coordinate system, by using two CCD cameras. The necessary distances between the CCD detector and the point in question are determined from adequate elements measured in the linear detector by using derived formulae. The size of physical surfaces (products) examined could be up to some tens of meters using this technique. The accuracy of the suggested method is analyzed below.

Optical digital levels [41] used in geodesy are based on the data received from a bar code on a staff and processing it in an optical-electronic measuring instrument. A view of the bars is projected onto a photocell matrix or a CCD camera sensor where the distance to the staff is assessed by focusing the optical image. Then a correlation of the bar code with the length pattern in the sensor is calculated. The bar images are processed into voltages and, subsequently, digital output. Most digital levels operate in such a mode. The view of code marks in the photocells is transformed into the electrical digital signals and using this information the distance and height are calculated by a microprocessor determining the correlation value. Three types of view development are used in modern level instruments: namely the correlation method, the geometrical method, and the phase shift method.

The geometric errors in the volume of a multicoordinate machine consist of the perpendicularity of coordinate axes $\Delta_{x/y}$, $\Delta_{x/z}$, $\Delta_{y/z}$; the coordinate position errors $\Delta_{x,y,z}$ along the axes x, y, and z; rolling errors $\Delta\varphi_{x,y,z}$ around the axes x, y and z; pitch and yaw errors $\Delta_{x(y,z)}$; during the movement of the part along the relevant axis in the indicated plane (Figure 6.1). In total there are 21 types of geometric errors of multicoordinate machines. Specific errors are present in the signal output from the measuring head of a CMM, and they are usually analyzed separately.

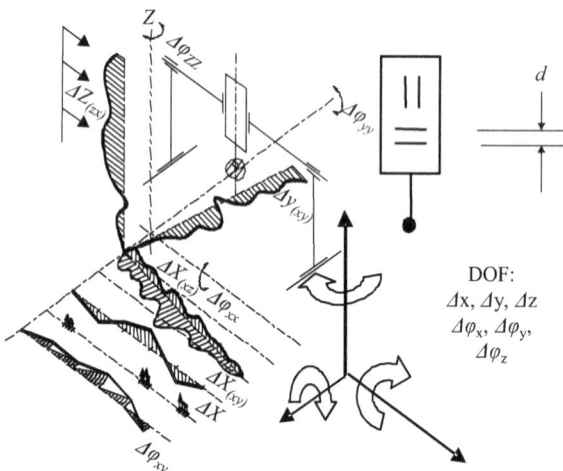

Figure 6.1. Diagram of the distribution of geometrical parameters' errors of CMM.

On the right-hand side of Figure 6.1 there are six degrees of freedom (DOF) depicted for the position of the MP in the same coordinate system of the machine [42]. For a full determination of the position of a body in space, all six DOF of the body must be determined. In fact, it would be possible to track bars placed on the body of the MP as it is utilized in the equipment mentioned above. Horizontal bars are used for vertical position determination and vertical bars for horizontal position determination. Nevertheless, additional measures must be undertaken for determining rolling, pitch, and yaw errors.

The distance S from the point of view to the bars is calculated as

$$S = \frac{d}{\tan(\alpha + \alpha') - \tan \alpha} \tag{6.1}$$

where d is the distance between the bars on the MP (Figure 6.1) calibrated to high accuracy. The angles α and α' are determined during the calibration of the detecting system.

A more thorough analysis of the accuracy of such a detection system is presented in a paper by Skeival as and Giniot is [43]. According to this analysis it can be stated that the standard deviation in detecting the position of the MP could be achieved within a range of several micrometers over a length of one meter.

6.3 VOLUMETRIC DETECTION OF POSITION

Volumetric detection of the object's position can be accomplished by using rotary encoders in the visual detection system. The measuring system containing the rotary encoders is placed at point A at some distance $L = OA$ from the tracking point (Figure 6.2).

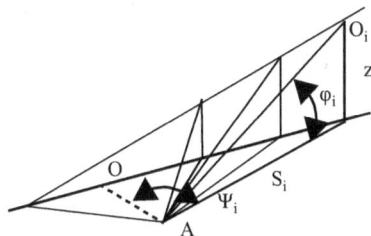

Figure 6.2. Measurement of the position of the object by angular measuring system.

There is enough data for the calculation of output data, the height of the object (MP) zi, that will take into the account the systematic errors of the basic value L and of the readings of angles φ_i and ψ_i:

$$Z_i = \frac{(L \pm \delta L) tg(\varphi_i \pm \delta \varphi_i)}{\cos(\psi_i \pm \delta \psi_i)}. \qquad (6.2)$$

where $i = 1,2,3,\dots$, and the distance S_i is given by

$$S_i = \frac{(L \pm \delta L) tg(\varphi_i \pm \delta \varphi_i)}{\cos(\psi_i \pm \delta \psi_i) tg(\varphi_i \pm \delta \varphi_i)}. \qquad (6.3)$$

Angles φ_I and ψ_I are measured by a rotary encoder fixed in the measuring system. Many angular and linear measurement transducers or encoders are used in industry and machine engineering for position and displacement measurements. The accuracy of angular position using these devices reaches approximately 0.3" to 0.1" (seconds of arc). An arrangement of measuring equipment consists of a digitizing camera and two rotary encoders for angular position measurement with registration devices placed at point A. The main parts of the device are the digital camera, a rotary encoder for angle ψ_i control, a rotary encoder for angle φ_i control, a trigger interface, indicating units and a registering device. The camera must be fixed to the gimbals with the axis connected to the rotary encoder for angle ψ_i control (measurement of angle in the horizontal plane), and the axis connected to the rotary encoder for measurement of angle φ_i in the perpendicular plane. During the measurements the readings of angles φ_i and ψ_i at the known distance L from the axis of the object movement are read for calculations. The technique proposed here is to be utilized in the CMM calibration process and Figure 6.3 shows a schematic diagram of the proposed system.

The idea is to place coded bars on the surface of a cylinder which is used instead of the measuring head of the CMM during performance verification. The cylindrical bar coded measuring head (BCMH) is fixed into the ram of the CMM. The CCD camera is placed on the base plate of the CMM and is targeted to the bars of the cylinder. The bars are marked on the surface of the cylinder in the vertical direction and around the cylinder. This provides the opportunity to determine distances in the vertical and longitudinal directions and the rotational movement of the ram about the vertical axis (yaw). During the detection of the BCMH position the CCD camera triggers the transfer of the digital values of the coordinates X, Y, and Z. The vertical position of the BCMH is determined by calculating

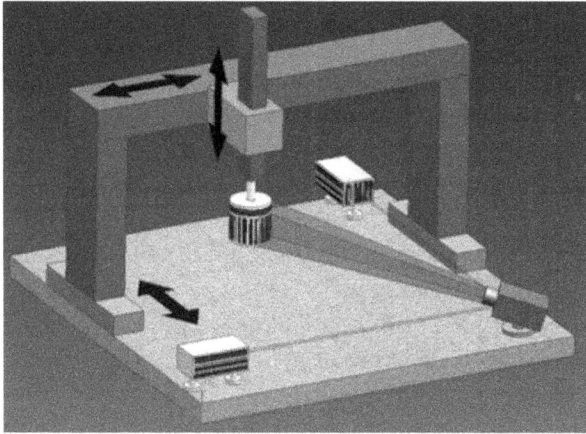

Figure 6.3. Measurement of the probe position in vertical and horizontal planes using coded bar.

the known (calibrated before) distance d between two circular lines on the BCMH in relation to the horizontal sight view of the camera. The position in the horizontal plane is determined by calculations of the displacement of the circular bar code marks placed on the BCMH. Additional information for these calculations is supplied by the horizontal and vertical reference bars (also calibrated) placed on the reference marks (RM) which are also shown in Figure 6.3. The problem remains to achieve a higher accuracy of angular readings.

The main aim of this proposal is to provide SMEs with the opportunity to acquire CMM performance verification data using an easily accessible instrument. This eliminates the need to use complicated measuring equipment together with highly qualified staff. Processing and evaluation of the data received by this method could be at a remote metrology center supplied with relevant software.

6.4 A PROPOSED TELEMETRIC CMM

The originality in this development project will be to produce a CMM design whereby the system that positions the probe is entirely independent from the system that calculates the location of the probe contact point as a 3D coordinate measurement. Both the development of a video-telemetry probe location system based on noncontact principles and a new light-weight positioning system will be innovative. This method is realizable for the determination of the coordinates of an object's surface points using

CCD (coupled charge device) cameras, the sensor of which consists of a matrix of photodiodes with a focal lens. Trigonometric calculations of distances and angles from two CCD cameras are used for the determination of a point's coordinates in a free coordinate system. For this purpose, a baseline between two CCD cameras is established. Figure 6.4 shows an alternative structure of the CMM, a telemetric coordinate measuring machine (TCMM).

The TMM has a base plate Q, on the surface of which all equipment is placed. A MP is fixed on the stand (ST). The coded bars are placed on the surface of the stand (ST) which is moved on the plate (Q) thus restricting at least two DOF. The movement can be accomplished by an air cushion and it can be driven by hand or a simplified drive. Three CCD cameras are shown, whereas the cameras CCD1 and CCD2 control the position in the horizontal plane, and the camera CCD3 controls the position in the vertical plane. It would also be possible to implant a linear translational transducer into the stand (ST) and two-directional angle sensors into the body of the MP. The readings from the autonomic transducer and sensors, together with the MP touching signal, would be transmitted by the information transmitter (IT), radio or Bluetooth type, fixed to the body of the ST.

The information on the position of the MP at the moment of touching the part to be measured is input into the control unit for the calculation of the coordinates of the points already measured [44]. It must be noted that there are additional requirements for ensuring the parameters of linear displacement accuracy when adding some additional DOF to the ST. Important parameters of angular DOF can be overcome using the coded bars in two planes in the volume.

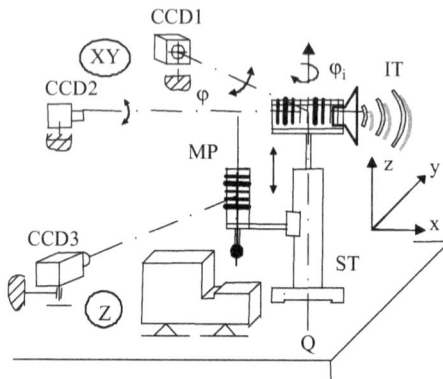

Figure 6.4. A telemetric coordinate measuring machine (TCMM).

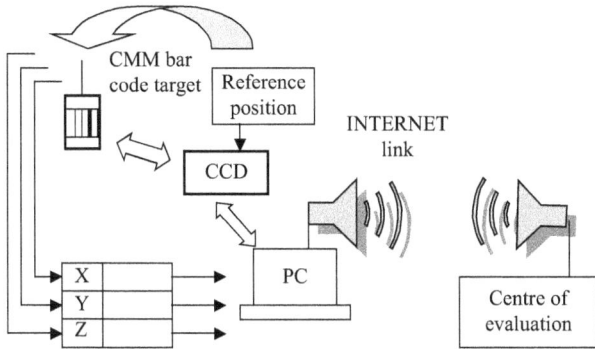

Figure 6.5. Generalized diagram of alternative 3D coordinate measuring system.

Figure 6.5 shows a generalized diagram of a proposed alternative 3D coordinate measuring system with an internet link to a remote evaluation center. Hence the processing and evaluation of the data received from this coordinate measuring system would not be undertaken on site but at the remote metrology center which would carry all the relevant software.

A feasible aim, to design a new type of CMM that improves the quality of measurement by eradicating the link between machine distortions and transducer errors, has been discussed. It is proposed to achieve this by separating the system that determines the probe location from the mechanism that actually positions the probe. This would lead to the production of a cheaper CMM. The next step would be to produce software algorithms capable of controlling the probe head position drive mechanism more accurately. Assessment of the 3D position of the object using an optical image transfer into the CCD camera can be accomplished by one of the methods used in geodetic instruments, that is by the correlation method, the geometrical method, or the phase shift method.

REFERENCES

1. Hunt, T.M. series ed. *The Concise Encyclopaedia of Condition Monitoring.* 1st ed. Oxford: Coxmoor Publishing Company, 2006.
2. Reeves, C.W. *The Vibration Monitoring Handbook*, ed. T.M. Hunt. 1st ed. Oxford: Coxmoor Publishing Company, Machine & Systems Condition Monitoring Series, 1998.
3. Wang, L.; and A.D. Hope. "Bearing Fault Diagnosis Using Multi-Layer Neural Networks." *Insight–Non-Destructive Testing and Condition Monitoring* 46, no. 8 (2004), pp. 451–5.
4. Wunderli, S. "Uncertainty and Sampling." *Accreditation and Quality Assurance* 8, no. 2 (February 2003), p. 90.
5. Ramsey, M.H. "Appropriate Rather than Representative Sampling Based on Acceptable Levels of Uncertainty." *Accreditation and Quality Assurance* 7, no. 7 (July 2002), pp. 274–80.
6. Chan, F.M.M.; T.G. King; and K.J. Stout. "The Influence of Sampling Strategy on a Circular Feature in Coordinate Measurements." *Measurement* 19, no. 2 (October 1996), pp. 73–81.
7. Lee, G.; J. Mou; and Y. Shen. "Sampling Strategy Design for Dimensional Measurement of Geometric Features Using Coordinate Measuring Machines." *International Journal of Machine Tools and Manufacture* 37, no. 7 (July 1997), pp. 917–34.
8. Woo, T.C.; and R. Liang. "Dimensional Measurement of Surfaces and their Sampling." *Computer Aided Design* 25, no. 4 (April 1993), pp. 233–9.
9. Hammersley, J.M. "Monte Carlo Methods for Solving Multivariate Problems." *Annals of the New York Academy of Sciences* 86, no. 3 (May 1960), pp. 844–74.
10. Bong, E.T.; and J.W. Han. "A Precision Length Measuring System for a Variety of Linear Artefacts." *Measurement Science and Technology* 12, no. 6 (June 2001), pp. 698–701.
11. Sladek, J.; and M. Krawczyk. "Modeling and Assessment of Large Cmms' Accuracy." *Proc. XVII IMEKO World Congress,* Dubrovnik, Croatia, 1903–1906, June 22–27, 2003.
12. Seung-Woo, K.; R. Hyug–Gyo; and J-Y. Chu. "Volumetric Phase—Measuring Interferometer for Three-Dimensional Coordinate Metrology." *Precision Engineering* 27, no. 2 (April 2003), pp. 205–15.

13. ISO/IEC Guide 98-1. *Uncertainty of Measurement—Part 1: Introduction to the Expression of Uncertainty in Measurement*, 2009.

14. EN ISO 10360-2. *Geometrical Product Specifications—Acceptance and Reverification Tests for Coordinate Measuring Machines (CMMs)*, 2009.

15. ASME B89.1.12M-1990. *Methods for Performance Evaluation of Coordinate Measuring Machines*, 1990.

16. CMMA. *Coordinate Measuring Machine Manufacturers Association. Accuracy Specification for CMMs*, 1991.

17. VDI/VDE 2617 Blatt 4/Part 4. *Verein DeutscherIngenieure*. Düsseldorf, 1989.

18. Giniotis, V.; M. Rybokas; and P. Petroškevičius. "Investigations into the Accuracy of Angle Calibration." *Geodesy and Cartography* 30, no. 3 (May 2004), pp. 65–70.

19. Giniotis, V. "Brief Review of Methods for Measuring of Circular Scales." *Geodesy and Cartography (Geodezijairkartografija)* 23, no. 2 (1997), pp. 21–5.

20. Giniotis V.; and G. Murauskas. "Development of Methods for Intelligent Measurement of the Raster Scales." In *Proceedings of the XIV IMEKO World Congress*, ed. J. Halttunen, Vol. 8, 234–9.Tampere, Finland, June 1–6, 1997.

21. Giniotis V., 1992. "Method of Measuring the Pitch Accuracy of Raster Scales." In *Risk Minimisation by Experimental Mechanics*, ed. VDI Berichte, Series 940, 41–5, 1992.

22. Giniotis V. Method of measurement of angular error of circular scale. SU Patent 1654644, Bull. No 21. (In Russian), 1991.

23. Giniotis, V.; and K.T.V. Grattan. "Optical Method for the Calibration of Raster Scales." *Measurement* 32, no. 1 (2002), pp. 23–9.

24. ISO 17123-3. *Optics and Optical Instruments—Field Procedures for Testing Geodetic and Surveying Instruments— Part 3: Theodolites*, 2001.

25. ISO 17123-5. Optics and Optical Instruments—Field Procedures for Testing Geodetic and Surveying Instruments—Part 5: Electronic Tacheometers, 2001.

26. Ingens and, H. "A New Method of Theodolite Calibration." *XIX International Congress*, 91–100, Helsinki, Finland, 1990.

27. Bansevičius, R.; and R.T. Tolocka. "Piezoelectric Actuators." In *The Mechatronics Handbook*, ed. R.H. Bishop, 51–62.Boca Raton, London: CRC Press, 2002.

28. ISO 9001. *Requirements for a Functioning Quality Management System*, 2008.

29. ISO 14405-1. *Geometrical Product Specifications—Dimensional Tolerancing—Part I, Linear Sizes*, 2010.

30. ISO 14405-2. *Geometrical Product Specifications—Dimensional Tolerancing—Part 2, Dimensions Other Than Linear Sizes*, 2011.

31. BS 3800–2. *Statistical Methods for the Determination of Accuracy and Repeatability of Machine Tools*, 1991.

32. Crawley, E.F.; and J. De Luis. "Use of Piezoelectric Actuators as Elements of Intelligent Structures." *AIAAJ* 25, no. 10 (October 1987), pp. 1373–85.

33. Bansevičius, R.; and V. Giniotis. "Mechatronic Correctional Devices for Precision Engineering. Mechatronic Systems and Materials." *Solid State Phenomena*113, MSM2005 (2006), pp. 429–34. Eds. N. Bizys; and A. H. Marcinkevicius.

34. Leopold, J.; H. Günther; and R. Leopold. "New Developments in Fast 3D-Surface Quality Control." *Measurement* 33, no. 2 (2003), pp. 179–187.

35. www.gom.com

36. Sladek, J.; and M. Krawczyk. "Modeling and Assessment of Large Cmms' Accuracy.*" Proceedings, XVII IMEKO World Congress* 1903–6. Dubrovnik, Croatia, 2003.

37. Wozniak, A.; and M. Dobosz. "CMM Probe Testing by Means of a Low Force Sensor." *Proceedings of the XVI IMEKO World Congress*, ed. M.N. Durak-basa et al, Vol. 8, 341–4. Vienna, Austria, 2000.

38. www.leica-geosystems.com

39. www.faro.com

40. www.aicon.de

41. NA 2002; and NA 3003. "Technical Report, Digital Levels." *Leica Geosystems,* 13. AG, Geodesy, 1997.

42. Giniotis V.; R. Bansevičius; And J.A.G. Knight. "Complex Accuracy Assessment of Multicoordinate Machines." *Proceedings of the 6th ISMQC IMEKO Symposium Wien, Metrology for Quality Control in Production*, ed. P.H. Osanna et al.,195–9, 1998.

43. Skeivalas, J.; and V. Giniotis. "Accuracy and Performance Analysis of Digital Levels." *Optical Engineering* 44, no. 2 (2005), pp. 027007-1-6.

44. Giniotis, V. *Position and Displacement Measurement.* Vilnius: Technika, 2005.

BIBLIOGRAPHY

Abbe, M.; and M. Sawabe. "Geometric Calibration of CMM by Means of 3-Dimensional Coordinate Comparison." In *Proceedings of the 6th ISMQC IMEKO Symposium*, ed. Osanna et al., 3–8, Vienna, Austria, 1998.

Bansevičius, R.; and J.A.G. Knight. "Intelligent Mechanisms with Piezoactive Links: State of the Art, Problems, Future Developments." *Proceedings of 10th World Congress on the Theory of Machines and Mechanisms*, 2008–2013, Onlu, Finland, 1999.

Bansevičius, R.; and R.T. Tolocka. "Piezoelectric Actuators." In *The Mechatronics Handbook*, ed. R.H. Bishop, 51–62. Boca Raton, London: CRC Press, 2002.

Bansevičius, R.; and V. Giniotis. "Mechatronic Means for Machine Accuracy Improvement." *Mechatronics* 12, no. 9–10 (November–December), pp. 1133–43, 2002.

Bansevičius, R.; R. Parkin; A. Jebb; and J. Knight. "Piezomechanics as a Sub-system of Mechatronics: Present state of the Art, Problems, Future Developments." *IEEE Transactions on Industrial Electronics* 43, no. 1 (1996), pp. 23–30.

Bansevičius, R.; V. Giniotis; A. Augustaitis; and A. Jurkauskas. Means for Determination of Accuracy of the Displacement for Automatic Machines. Patent SU No 1705699, CL. GO1B 11/00, Bull. No 2 (in Russian), 1992.

Bručas, D.; V. Giniotis; and P. Petroškevičius. "The Construction of the Test Bench for Calibration of Geodetic Instruments." *Geodezija ir kartografija (Geodesy and Cartography)* 32, no. 3 (2006), pp. 66–70.

D'Hooghe, F.P.; P. Schellekens; and L. Levasier. "Geometric Calibration of CMMs Using 3D Length Measurements." *Proceedings of the XVI IMEKO World Congress*, ed. M.N. Durakbasa et al., Vol. 8, 121–6. Vienna, Austria, 2000.

Downs, M.J.; A.B. Forbes; and J.E. Siddle. "The Verification of A High-Precision Two-Dimensional Position Measurement System." *Measurement Science and Technology* 9, no. 7, pp. 1111–4, 1998.

Gandhi, M.V.; and B.S. Thompson. *Smart Materials and Structures*. Chapman & Hall, 1992.

Giniotis, V. "Position and Displacement Measurement." In *Padětiesirposlinkiųmatavimas*, 215. Vilnius: Technika, 2005.

Giniotis, V. Method of Measurement of Angular Error of Circular Scale. SU Patent, No 1654644, Bull. No 21. (In Russian), 1991.

Giniotis, V.; and K.T.V. Grattan. "Optical Method for the Calibration of Raster Scales." *Measurement* 32, no.1 (2002), pp. 23–29.

Giniotis, V.; and M. Rybokas. "Data Processing and Information Assessment in Scales Measurement Simulation." *XVII IMEKO World Congress*, 1053–6, 2003.

Giniotis, V.; K.T.V. Grattan; M. Rybokas; and D. Brucas. "Analysis of Measurement System as the Mechatronics System." *Journal of Physics Conference Series* 238, no. 1 (2010), 012021.http://iopscience.iop.org/1742-6596/238/1/012021

Giniotis, V.; K.T.V. Grattan; M. Rybokas; and R. Kulvietiene. "Uncertainty and Indeterminacy of Measurement Data." *Measurement* 36, no 2 (2004), pp. 195–202.

Giniotis, V.; M. Rybokas; and D. Brucas. "Application of Mechatronic Means for Precision Measurements//Mechatronics." *Proceedings*, June 28–30, Swiss Federal Institute of Technology ETH, Zurich, Switzerland, Book 2, 250–6, 2010.

Giniotis, V.; R. Bansevičius; and J.A.G. Knight. "Complex Accuracy Assessment of Multicoordinate Machines." *Proceedings of the 6th ISMQC IMEKO Symposium Wien, Metrology for Quality Control in Production,* ed. P.H. Osanna et al., 195–9, 1998.

Grattan, K.T.V.; and B.T. Meggit. *Optical Fibre Sensor Technology 2: Devices &Technology*, 440. London: Chapman & Hall, 1998. (Guide to the expression of uncertainty in measurement, 2008, JCUM 100:2008 (GUM 1995 with minor corrections).

Humienny, Z. ed.; P.H. Osanna; M. Tamre; A. Weckenmann; L. Blunt; and W. Jakubiec. *Geometrical Product Specifications*. Course for Technical Universities, Warsaw University of Technology Printing House, Warsaw, 2001.

Just, A.; M. Krause; R. Probst; and R. Wittekopf. "Calibration of High-Resolution Electronic Autocollimators Against an Angle Comparator." *Metrologia* 40, no. 5 (October 2003), pp. 288–94.

Klein, M.; J. Eichenberger; and T. Delio. "Noncontact Vibration Monitoring for High-Speed Machine Tools." *Machining Technology* 16, no. 4 (October 2005), pp.1–5.

Lee, G.; J. Moum; and Y. Shen. "Sampling Strategy Design for Dimensional Measurement of Geometric Features Using Coordinate Measuring Machines." *International Journal of Machine Tools and Manufacture* 37, no. 7 (July 1997), pp. 917–34.

Leopold J.; H. Günther; and R. Leopold. "New Developments in Fast 3D-Surface Quality Control." *Measurement* 33, no. 2 (2003), pp. 179–87.

Marshall, R.H.; Y.N. Ning; A.W. Palmer; and K.T.V. Grattan. "Accurate Displacement Measurement Using a Novel Fibre-Optic Electronically Scanned White Light Interferometer." *Proceedings of the XIV IMEKO World Congress,* ed. J. Halttunen, Vol. 2, 144–9. Tampere, Finland, June 1–6, 1997.

Osawa, S.; T. Takatsuji; and T. Kurosawa. "Development of a New Artefact to Calibrate a Ball Plate." *Proceedings of the XVI IMEKO World Congress*, ed. M.N. Durakbasa et al., Vol. 8, 221–6, 2000.

Paakari, J.; and H. Ailisto. "On-Line Dimensional Quality Control Of Large Objects Using Moiré Contouring." *Proceedings of the XIV IMEKO World Congress,* ed. J. Halttunen, Vol. 8, 106–12. Tampere, Finland, June 1–6, 1997.

Probst, R. "Requirements and Recent Developments in High Precision Angle Metrology." *The 186th PTB-Seminar,* 124–31, 2003.

Rybokas, M.; V. Giniotis; P. Petroškevičius; R. Kulvietienė; and D. Bručas. "Performance and Accuracy Monitoring of Geodetic Instruments." *Proceedings of the International Conference on Condition Monitoring,* 167–72. Kings College, Cambridge, UK, July 18–21, 2005.

Toyama, S.; S. Sugitani; Z. Gnogiang; Y, Miytani; and K. Nakamura. "Multi-Degree-of-Freedom Spherical Ultrasonic Motor." *Proceedings of the IEEE International Conference on Robotics and Automation*, Vol. 3, 2935–40, 1995.

Tyler, E.W. et al. "Practical Aspects of Touch-Trigger Probe Error Compensation." *Precision Engineering* 21, no. 1 (1997), pp. 1–17.

Weekers, W.G.; and P.H. Schellekens. "Compensation for Dynamic Errors of Coordinate Measuring Machines." *Measurement* 20, no. 3 (1997), pp. 197–209.

Wozniak, A.; and M. Dobosz. "CMM Probe Testing by Means of a Low Force Sensor." *Proceedings of the XVI IMEKO World Congress*, ed. M.N. Durakbasa et al., Vol. 8, 341–4, 2000.

Yandayan, T. "Application of the Novel Technique for Calibration of 23 Sided Polygon Angles With Non-Integer Subdivision of Indexing Table." *8th Intern Symposium on Measurement, Quality Control in Production* 769–74. Düsseldorf, October 12–15, 2004.

INDEX

THIS TITLE IS FROM OUR AUTOMATION AND CONTROL AND MECHANICAL ENGINEERING COLLECTION. MORE TITLES THAT MAY BE OF INTEREST INCLUDE...

Alarm Management for Process Control: A Best-Practice Guide for Design, Implementation, and Use of Industrial Alarm Systems
By Douglas H. Rothenberg

Advanced Regulatory Control: Applications and Techniques
By David W. Spitzer

Process Control Case Histories: An Insightful and Humorous Perspective from the Control Room
By Gregory K. McMillan

THE WBF BOOK SERIES which includes: ISA 88 Implementation Experiences, Applying ISA 88 In Discrete and Continuous Manufacturing, ISA 95 Implementation Experiences, and ISA 88 and ISA 95 in the Life Science

Going the Distance: Solids Level Measurement with Radar
By Tim Little, Henry Vandelinde

Catching the Process Fieldbus: An Introduction to PROFIBUS for Process Automation
By James Powell, Henry Vandelinde

Plant IT: Integrating Information Technology into Automated Manufacturing
By Dennis L. Brandl, Donald E. Brandl

Automated Weighing Technology: Process Solutions
By Ralph Closs, Henry Vandelinde, Matt Morrissey

Automating Manufacturing Operations: The Penultimate Approach
By William M. Hawkins

Flexible Test Automation: A Software Framework for Easily Developing Measurement Applications (forthcoming!)
By Pasquale Arpaia, Vitaliano Inglese, Ernesto De Matteis

Situation Management: Situation Awareness and Decision-making for Operators in Industrial Control Rooms and Operation Centers (forthcoming!)
By Douglas H. Rothenberg, Ian Nimmo

Tuning and Control Loop Performance, 4e (forthcoming!)
By Gregory K. McMillan

Announcing Digital Content Crafted by Librarians

Momentum Press offers digital content as authoritative treatments of advanced engineering topics, by leaders in their fields. Hosted on ebrary, MP provides practitioners, researchers, faculty and students in engineering, science and industry with innovative electronic content in sensors and controls engineering, advanced energy engineering, manufacturing, and materials science. **Momentum Press offers library-friendly terms:**

- perpetual access for a one-time fee
- no subscriptions or access fees required
- unlimited concurrent usage permitted
- downloadable PDFs provided
- free MARC records included
- free trials

The **Momentum Press** digital library is very affordable, with no obligation to buy in future years.

For more information, please visit **www.momentumpress.net/library** or to set up a trial in the US, please contact **mpsales@globalepress.com**.

www.ingramcontent.com/pod-product-compliance
Lightning Source LLC
Chambersburg PA
CBHW070715220326
41598CB00024BA/3161